Monographs on Endocrinology

Volume 24

J. C. Brown

Gastric Inhibitory Polypeptide

With 32 Figures and 1 Table

Springer-Verlag
Berlin Heidelberg New York 1982

John Canvin Brown

Department of Physiology,
University of British Columbia,
Vancouver, B.C., Canada

ISBN 3-540-11271-5 Springer-Verlag Berlin Heidelberg New York
ISBN 0-387-11271-5 Springer-Verlag New York Heidelberg Berlin

Library of Congress Cataloging in Publication Data.

Brown, J.C. (John Canvin), 1938– . Gastric inhibitory polypeptide.
(Monographs on endocrinology; v. 24)
Bibliography: p. Includes index.
1. Gastrointestinal hormones. I. Title. II. Series.
[DNLM: 1. Gastrointestinal hormones. W1 MO57 v. 24/WK 170 B878g]
QP572.G35B76 1982 612′.32 81-23301
ISBN 0-387-11271-5 (U.S.) AACR2

Printing and binding: Oscar Brandstetter Druckerei GmbH & Co. KG, Wiesbaden
2127/3020-543210

To: D. Harold Copp

(C.C., B.A., M.D., Ph.D., LL.D., D.Sc., F.R.S.C., F.R.S.)

Head of the Department of Physiology,
University of British Columbia,
Vancouver, B.C., Canada,
from July 1950 to June 1980

Preface

Dr. Raymond Pederson, Dr. Jill Dryburgh and I commenced work on GIP in 1968, when, with the generous help of Professor Viktor Mutt and Professor Erik Jorpes of the Karolinska Institute, Stockholm, we were able to establish that there existed an inhibitory material for acid secretion in cholecystokinin-pancreozymin preparations. Once the physiological evidence for the inhibitor was established it seemed appropriate to seek help in its isolation. Dr. J. Dryburgh and Dr. R. Pederson were left to bioassay fractions in Vancouver whilst I enjoyed the company of Professor Mutt at the Karolinska for one year, as a Medical Research Council of Canada Visiting Scientist. Purification of the inhibitory factor proceeded rapidly due, in no small measure, to Professor Mutt's untiring efforts on my behalf. Later that year, Dr. Dryburgh joined us in Stockholm to begin the sequence work on GIP. This was completed late in 1970 in Vancouver.

In Stockholm in June 1970, I met a fellow Canadian Dr. John Dupré (McGill University) at a cocktail party who kept commenting about the possibility of GIP being an insulinotropic hormone, the "incretin" of earlier days. At that time, gastrointestinal physiologist as I was, I did not recognize the importance of his comment. This became apparent two or three years later when Dr. Dupré demonstrated that GIP was insulinotropic in man.

In 1972, Maryanne Kuzio and Dr. Kathleen Malloy joined us and before I was aware of what was happening, we had something called a radioimmunoassay to GIP. Dr. Samuel Cataland (Ohio State University) began collaborating with us at this time and eventually introduced us all to Dr. Thomas O'Dorisio, but still we had only one antiserum for GIP. The radioimmunoassay certainly helped with the elucidation of some of the physiology of GIP and also was useful in investigating the possible pathophysiological role for GIP.

In 1973 I met Professor Werner Creutzfeldt (University of Göttingen) at a meeting in Davos, Switzerland, and realized that Göttingen would be an excellent place to establish the RIA for GIP to investigate possible clinical states in which GIP might be involved. I spent one year (1974–75) in Professor Creutzfeldt's department and actually became familiar with the multiple interpretations of the endocrine disorder known as diabetes. Whether or not GIP has a role in pathophysiological states is equivocal.

A few apologies are in order; in particular to the synthetic chemists, Dr. Jack Morley and colleagues, Professor Erich Wünsch and colleagues, and Professor Noboru Yanaihara and colleagues for synthesizing to an incorrect sequence. Dr. Hans Jörnvall and Professor Mutt have I hope finally corrected the sequence and if an acceptable interpretation of the acronym GIP could be settled upon, then at least one chapter would be completed.

My gratitude to my current colleagues and associates, Sam Kwauk, Marshall Dahl, Susan Otte, Christopher McIntosh and Raymond Pederson whose contributions to the GIP saga, past and present, are of immeasurable significance.

This monograph on GIP had a "gestation" period of six years, but the actual period of "labour" was relatively short. I am certain that its rapid delivery will have resulted in at least minor errors and perhaps irritating omissions. I accept sole responsibility for these.

Vancouver, January 1982 J.C. Brown

Contents

A. Introduction

I. Enterogastrone Concept

In the study of the humorally mediated gastric inhibitory mechanism or mechanisms, originating from the small intestine, two avenues of research have been developed. Physiological stimuli, e.g. fat, acid and hypertonic solutions, have been introduced into the small intestine and the spectrum and potency of their inhibitory actions on gastric physiology, in innervated and denervated preparations, observed. The second approach has been via attempts to isolate and purify the substances from the gastro-intestinal mucosa which produce the inhibitory effects.

1. Endogenous

The term enterogastrone was originally coined by Kosaka and Lim (1930) to describe a hypothetical inhibitory hormone for gastric acid secretion which became liberated from the mucosa of the upper small intestine by the presence of fat or the digestion products of fats, the fatty acids. Ewald and Boas (1886) first recognized that the addition of olive oil to a test meal of starch paste produced a delay in gastric emptying and an inhibition in gastric secretion when given to human subjects. Pavlov (1910), using Pavlov fundic pouches, was able to demonstrate that addition of fat to a meal fed to dogs inhibited secretion of acid and pepsin, and further, that the reflex secretion of acid and pepsin produced by sham feeding was inhibited by prior feeding with olive oil. The same treatment inhibited motor activity in the stomach and brought about a delay in gastric emptying, and it was the presence of fat in the duodenum rather than in the stomach itself which produced this inhibitory action. Transplanted gastric pouches were used in experiments by Farrell and Ivy (1926) in which they demonstrated that fat inhibited gastric motor activity and that a humoral agent was involved. Inhibition of gastric secretion from transplanted pouches by a similar mechanism was demonstrated by Feng et al. (1929). They excluded the possibility that either absorbed fat and its digestion products, or the passage of bile into the intestine following fat ingestion was producing the inhibitory effect.

Sircus (1958) infused olive oil into an isolated duodenal loop and was able to demonstrate the necessity for the fat to be in an absorbable form to produce inhibition of acid secretion. The oil had to be pre-incubated with pancreatic juice. Exclusion of bile and pancreatic juice, which would also interfere with fat absorption, produced a lesser degree of inhibition (Menguy 1960). A correlation between the duration of inhibition and the rate of absorption was shown by Long and Brooks (1965) using ^{14}C oleic acid and triolein. Duration of inhibition was longer and absorption slower with oleic acid than with triolein. Further support for an interrelationship between absorption and inhibition was supplied by Konturek and Grossman (1965), who perfused isolated intestinal loops with micellar fat mixtures. They found that the degree of inhibition was dependent on the rate of absorption and that inhibition could be produced from all levels of the small intestine. However, the degree of inhibition was greatest from jejunal loops, where the rate of fat absorption was fastest.

Fat and the fatty acids in the small intestine are not the only initiators of mechanisms for inhibition of acid secretion and motor activity. Hypertonic sugar solutions introduced into the duodenum in conscious dogs were demonstrated by Quigley and Phelps (1934) to inhibit the motor activity of transplanted fundic pouches, and it was shown that the effect was not due to glucose absorption. Day and Komarov (1939) were able to inhibit acid secretion stimulated by sham feeding in dogs by introducing glucose solutions with concentrations of above 20% into the duodenum. However, only slight inhibition was observed when the stimulus was histamine. They excluded the possibility that inhibition was produced as the result of hyperglycaemia following glucose absorption by showing that intravenous injection of glucose produced insignificant inhibition. Hypertonicity in the duodenum produced by other sugars, polysaccharides, saline and peptone solutions suggested the existence of a common osmoreceptor mechanism sensitive to changes in osmolarity, but Sircus (1958) demonstrated that intraduodenal hypertonicity produced inhibition in transplanted fundic pouches, reaffirming a humoral mechanism for inhibition. Konturek and Grossman (1965) were able to show that the degree of inhibition produced by a 20% glucose solution in the duodenum was less than the effect observed with fats.

The first demonstration that acidification of the small intestine would inhibit gastric secretion was provided by Sokolov (1904, cited by Babkin 1944), when he showed that introduction of 0.5% hydrochloric acid into the duodenum markedly diminished acid secretion from a Pavlov pouch which had been stimulated by a meat meal. Inhibition of acid secretion stimulated by sham-feeding was shown by Day and Webster (1935) and Pincus et al. (1942) reported that the duodenal pH must be as low as 2.5 before acid secretion is inhibited. Andersson (1960 a, b, c) established that acid in the duodenum produced inhibition of gastric secretion via a humoral mechanism. Acidification of the duodenum of dogs inhibited acid secretion both in vagally innervated and vagally denervated gastric pouches during fasting (Andersson 1960a) and in response to a meal (Andersson 1960b). The fact that acid secretion in response to exogenous gastrin was also inhibited by duodenal acidification suggested that the humoral agent produced inhibition at some point after the release of gastrin (Wormsley and Grossman 1964). Andersson (1960c), Andersson and Grossman (1965) and Johnson and Grossman (1968) demonstrated that duodenal acidification did not inhibit histamine-stimulated secretion in the dog. The effectiveness of inhibitory mechanisms initiated by duodenal acidification against histamine–induced acid secretion has been described as significant by Code and Watkinson (1955) and by Wormsley and Grossman (1964), but as insignificant by Andersson et al. (1965). It is probable that a variety of factors influence the inhibitory mechanism to histamine–stimulated gastric secretion. These include (a) the interaction between endogenous release of gastrin and injected histamine, (b) the dose of histamine and type of preparation used and (c) whether the pouch or whole stomach was innervated or denervated. Andersson and Grossman (1965) were able to show in Heidenhain pouch dogs with vagally innervated antral pouches that the response to graded doses of histamine was significantly reduced by acidification of the antral pouch, suggesting that when the antral mucosa was removed from contact with acid there could be sufficient endogenous gastrin released to potentiate the stimulatory effect of injected histamine. Inhibition observed in such an experimental situation could be due to suppression of gastrin release.

A variety of substances will produce inhibition of gastric secretion when introduced into the duodenum. Acid secretion from a denervated fundic pouch stimulated by exogenous gastrin is more easily inhibited than acid secretion in a vagally innervated pouch or from the whole stomach or when histamine is the stimulus.

2. Exogenous

The acid responses of Heidenhain pouch animals to a meal and to histamine were shown by Kosaka and Lim (1930) to be inhibited by injection of saline extracts of the duodenal mucosa. Purification of the inhibitory extracts was attempted and they were reportedly free from secretin and vasodilator agents. However, it was shown that other tissue extracts also had inhibitory properties (Kosaka et al. 1932). The extracts were active against meal-stimulated and histamine-stimulated gastric secretion in doses of 3-4 mg/kg body weight, and inhibition of motor acitivity from gastric fistulae dogs (i.e. the innervated stomach) was also described. Gray et al. (1937) described the preparation of an enterogastrone extract from the duodenal mucosa of hogs which inhibited both hunger contractions and distension-induced contractions in the empty fasted stomach of dogs. The preparation was active whether given intravenously or subcutaneously. The preparation was also inhibitory for acid secretion from Pavlov pouches or gastric fistulae when secretion was stimulated by injection of histamine or by feeding a mixed meal. The extent of the inhibitory effects on secretion and motor activity was also shown to be dependent upon the dose administered. Further purification of their enterogastrone extract, involving (a) ethyl alcohol fractionation, (b) iso-electric precipitation at pH 8.4 or (c) pyridine, phenol or picric acid precipitation, proved unsatisfactory, and a later study by Greengard et al. (1946) also failed in this respect.

At the conclusion of the purification attempts, both secretin and cholecystokinin-pancreozymin (CCK-PZ) activity were invariably retained. Wormsley and Grossmann (1964), using an essentially pure preparation of secretin (Jorpes and Mutt 1961), demonstrated an inhibitory effect on acid secretion stimulated by exogenous gastrin. Confirmation of these studies was provided by Gillespie and Grossmann (1964), who also found that a CCK-PZ-containing preparation of Jorpes et al. (1964), a less pure preparation than the secretin, inhibited acid secretion from an Heidenhain pouch stimulated by exogenous gastrin.

This CCK-PZ-containing preparation also inhibited low dose histamine-stimulated acid secretion. Although inhibition of gastric acid secretion and motor activity by release of endogenous humoral agents had been well documented, the nature of the humoral agent was still unknown.

The first of the hormones from the duodenum, which inhibit gastric secretion, to have its structure elucidated was secretin (Mutt and Jorpes 1966). Total synthesis was completed in the same year. Bodansky et al. (1966) and Vagne et al. (1968) found that the inhibitory properties of the synthetic material were the same as those of natural secretin. Secretin, and not an impurity, was confirmed to be an inhibitor of acid secretion from Heidenhain pouches of dogs when gastrin was the stimulus for acid secretion. Secretin was, however, shown not to be an inhibitor of acid secretion in Pavlov pouch dogs in which acid secretion was stimulated by insulin hypoglycaemia or histamine (Way 1970).

The original observation of Gillespie and Grossman (1964) that an impure CCK-PZ preparation would inhibit gastrin-stimulated acid secretion was extended by Brown and Magee (1967), who demonstrated that an impure CCK-PZ preparation inhibited acid secretion from Heidenhain pouches which had been stimulated by endogenously released gastrin. The CCK preparation used in this study was approximately 10% pure, containing 250 Ivy dog units (IDU) of cholecystokinin activity. Small doses of the same CCK-PZ preparation had been found to be stimulatory for acid secretion when administered intravenously to dogs in the basal state (Preshaw and Grossman 1965; Murat and White 1966; Magee and Nakamura 1966). The latter workers suggested that CCK-PZ behaved like gastrin, acting as a stimulant for acid secretion when administered in small doses but as an inhibitor when given in large doses.

II. Incretin Concept

1. Endogenous

Unger and Eisentraut (1969) introduced the term "entero-insular axis" to describe a proposed regulatory mechanism in which the hormones of the gastro-intestinal tract exerted an influence upon the secretion of the pancreatic islet cell hormones. They suggested that if such an interrelationship between the gastro-intestinal tract and the pancreas existed, its function would be to augment or accelerate the islet cell hormone responses to the ingested substrates, so that when appropriate, insulin or glucagon or both would be quantitatively mobilized in response to signals from the gastrointestinal tract (initiated by substrate concentration acting on the mucosa) rather than arterial concentrations of absorbed substrates. A possible interrelationship between these two endocrine organs was alluded to in the early days of endocrinology, when Moore et al. (1906) indicated that oral administration of extracts made from porcine duodeno-jejunal mucosa, which contained secretin, improved the glycosuria of diabetic patients. Although these studies were not extended, indicating, perhaps, the inconclusiveness of the effect, they were instrumental in stimulating an interest in the hypoglycaemic properties of duodenal extracts (Dixon and Wadia 1926; Laughton and Macallum 1932).

The duodenal substance with hypoglycaemic activity has been variously designated as "hypoglycaemic secretin", "incretin", "duodenin" or "insulinotropic hormone". Evidence was presented by Zunz and Labarre (1929) and Labarre and Still (1930) that the hypoglycaemic factor was something other than secretin, and Labarre (1932) is credited with the introduction of the name "incretin" to describe the gastrointestinal factor. Loew, et al. (1940) failed to isolate a hypoglycaemic substance from the duodenum; however, in their assay system the numerous extracts were only tested in fasted animals and in no instance were any administered to animals in which a degree of hyperglycaemia had been induced. They also described that the presence of HCl in the duodenum lowered neither fasting nor hyperglycaemic blood sugar levels, and with extensive support from the literature demonstrated the considerable disagreement amongst workers on the hypoglycaemic properties of duodenal extracts. Their conclusion that the results obtained "denied the entire evidence which has been advanced in support of the theory that the duodenum exerts a hormonal control over carbohydrate metabolism by producing a hypoglycaemic substance" had a significant negative impact on this area of research.

The development of the technique of radioimmunoassay for insulin heralded a reappraisal of the incretin problem. Elrick et al. (1964) showed a significant difference in plasma insulin responses to oral and intravenous glucose, in that a greater and more sustained increase occurred following glucose administered orally. These workers compared the two routes of glucose administration in the same individual, used equal glucose loads and achieved similar blood glucose levels. They stated that the significantly greater insulin release with oral glucose was indicative of an additional stimulus for insulin secretion and suggested that a gastro-intestinal factor, released by the presence of glucose in the stomach or upper small intestine, might be involved. McIntyre et al. (1964) compared in man the effects of glucose administered intravenously or by direct jejunal infusion. Lower blood glucose levels were obtained during the jejunal infusion, but a higher plasma insulin response to glucose given jejunally than to glucose given by intravenous infusion was observed. They concluded from their observations that a humoral substance was released from the jejunal wall during glucose absorption and that this agent acted in conjunction with the rising blood glucose level to stimulate insulin release.

2. The Insulinotropic Effect of the Gastrointestinal Hormones

The anatomical positioning of the cells releasing the gastrointestinal hormones, i.e. exposed to the luminal digestive products and in close proximity to the nutrient absorptive sites, is suggestive of an important role for the hormones in the metabolism of the nutrients. Evidence presented to support a role for the gastrointestinal hormones in the entero-insular axis is equivocal. Several factors have contributed to this, including the absence of information concerning the chemical purity of the preparations used in earlier studies. Pharmacological doses of the preparations under investigation (administered as pulse injections) were often employed, and indeed in the cases of secretin and CCK-PZ accurate serum levels are still not available. Possible interreaction of hormone preparations with circulating levels of nutrients were not investigated or controlled. In later studies, after the introduction of the radioimmunoassay for insulin, reports have often neglected to describe the nature of the insulinotropic response observed. A total immunoreactive insulin (IRI) response would be described with no indication as to whether both phases of IRI release were affected of only one phase. Usually, only the initial phase of insulin release has been observed. The availability of purer preparations of secretin and CCK-PZ (Jorpes and Mutt 1966) and gastrin (Gregory and Tracy 1964) have allowed these preparations to be more carefully scrutinized for incretin activity.

a) Secretin

Dupré (1964) demonstrated that injection of a crude secretin preparation significantly increased the disappearance rate of intravenously administered glucose and elevated IRI in man. Similar preparations when administered as a large single pulse injection, stimulated insulin release both in vitro (Pfeiffer et al. 1965) and in vivo (Dupré et al. 1966). A greater insulinotropic action of secretin has been demonstrated during hyperglycaemia than in the fasting state, and it was suggested by Dupré et al. (1969) that IRI release by secretin was potentiated by glucose. Arginine produced a similar IRI response to secretin, when infused in the fasting state. More frequent sampling of blood in the first 20 min after secretin injection revealed that secretin had a more profound effect on the readily releasable insulin pool, and Lerner and Porte (1970) suggested that insulin release in response to secretin was from a small storage pool.

In a later study in man, Lerner and Porte (1972) compared the IRI response to repeated large doses of secretin (150 U) at 30-min intervals with the acute response to glucose injection. They observed that repeated injection of secretin elicited an acute insulin response which decreased in magnitude. The acute response to glucose, however, was increased following the secretin injections. When repeated injections of glucose were administered, the acute IRI response to glucose decreased in magnitude whilst the response to secretin was increased. They proposed that glucose and secretin probably stimulated functionally separated storage pools of readily releasable insulin.

Buchanan et al. (1968) indicated that the insulinotropic action of secretin might be pharmacological when they were unable to demonstrate increase in insulin secretion after administering secretin to anaesthetized dogs in doses sufficient to produce a copious flow of pancreatic juice. Earlier, Wang and Grossman (1951) had been unable to demonstrate an increase in exocrine secretion from the transplanted pancreas of the dog when the duodenum was perfused with glucose. Recent studies with radio-immunoassays for secretin have failed to demonstrate an elevation in serum IR-secretin following ingestion of a mixed meal (Bloom et al. 1975; Chey et al. 1975) or of glucose (Bloom 1974; Boden et al. 1975). The necessary criteria establishing secretin as an insulinotropic hormone with a role in the entero-insular axis have not been satisfied.

b) Gastrin

Like secretin, gastrin has not been shown to satisfy the necessary criteria for an insulinotropic hormone participating in the entero-insular axis. Insulin release in response to injected gastrin was monophasic, transitory and unaffected by the state of glycaemia in the experimental situation (Unger et al. 1967; Dupré et al. 1969; Creutzfeldt et al. 1970). In instances where pure exogenous gastrin has been shown to significantly stimulate insulin release, the required dose has been pharmacological. Plasma gastrin levels show only slight and transient elevation after oral glucose in man (Buchanan 1973). Conflicting reports on the effect of exogenous gastrin exist, both stimulation (Dupré et al. 1969; Iverson 1971; Unger et al. 1967) and no effect (Buchanan et al. 1969) having been reported. Marks and Turner (1977) considered that there was no evident correlation between the increase in endogenous plasma gastrin concentrations produced by appropriate stimulation in man and stimulation of insulin release as measured by an increase in plasma insulin concentrations.

c) Cholecystokinin-pancreozymin (CCK-PZ)

Dupré and Beck (1966) showed that an intestinal mucosal extract with biological activity similar to CCK-PZ enhanced the increase in serum IRI produced by the intravenous infusion of glucose. Under appropriate conditions it also stimulated the release of immunoreactive glucagon (Buchanan et al. 1968; Unger et al. 1967), and Füssganger et al. (1969), in studies using the isolated perfused rat pancreas, suggested that insulin release was secondary to glucagon release. In all the reported studies on the insulinotropic action of exogenous CCK-PZ, a partially purified porcine preparation produced by Jorpes and Mutt was used. It was from this preparation that GIP was initially isolated (Brown et al. 1969; Brown et al. 1970). The results obtained from studies involving the endogenous release of CCK-PZ offer little support for an insulinotropic role of CCK-PZ.

Bioassay techniques for the measurement of CCK-PZ have provided most of the data relating to its physiological role. Problems have been experienced in the development of a radioimmunoassay for CCK-PZ (Go and Reilly 1975), making a confident interpretation of radioimmunoassay results very difficult. Bioassay techniques have demonstrated that CCK-PZ release can be induced by the intraduodenal administration of protein hydrolysates and fats (Wang and Grossman 1951), an observation which was extended by Meyer (1975) to include observations on the secretagogue effect of individual amino acids. L-phenylalanine and L-tryptophan seem to be the amino acids which have been described to be most consistent in their ability to release CCK-PZ, as measured by gallbladder contraction or release of pancreatic enzymes (Wang and Grossman 1951; Go et al. 1970; Meyer and Grossman 1972; Meyer et al. 1973). Intraduodenal arginine infusion was observed to have only a weak stimulatory effect on CCK-PZ release by Konturek et al (1972). Harvey et al (1973) have confirmed these increases in immunoreactive CCK-PZ (IR-CCK-PZ) and also reported an increased release in response to oral glucose, an observation which was consistent with earlier reports.

On reviewing the literature, Marks and Turner (1977) concluded that CCK-PZ probably had some insulin-releasing effects in the presence of hyperaminoacidaemia and could be capable of potentiating the insulin-stimulating effect of hyperglycaemia, but that it is unlikely to be responsible for the insulinotropic effect of oral glucose. CCK-PZ was considered by these workers to make a contribution to the overall insulin response to a mixed meal containing fat, protein and carbohydrate. This response to a mixed meal has been shown to be greater than the sum of the responses to the individual constituents.

III. Evidence for the Existence of GIP

1. Physiological Studies

Brown and Pederson (1970), whilst engaged in a multiparameter study to investigate the actions of gastrointestinal extracts containing CCK-PZ, presented evidence that apart from CCK-PZ itself there was possibly another inhibitor of acid secretion in some of these preparations. The CCK-PZ preparations used in this study had been described as possessing inhibitory activity for exogenous gastrin-stimulated acid secretion (Gillespie and Grossman 1964) and endogenous gastrin-stimulated acid secretion (Brown and Magee 1967). Preparations of similar purity had also been described by Johnson and Magee (1965) as being inhibitory for basal motor activity and also inhibitory for stimulated motor activity (Brown et al. 1967). In apparent conflict with the described acid inhibitory effect of CCK-PZ, Magee and Nakamura (1966) and Murat and White (1966) described that in the basal situation the preparations stimulated acid secretion. Since these experimental observations were made with CCK-PZ preparations which were impure, it was suggested that the gastric effects could have resulted from the actions of factors other than the hormone CCK-PZ.

This hypothesis was tested by Brown and Pederson (1970). The approach employed was to compare the effects of two preparations of CCK-PZ on the gastric parameters of acid secretion, pepsin secretion and motor activity. These preparations contained approximately 200 IDU and 1500 IDU/mg and were described as being '10% pure' and '40% pure'.

The animal model used in these studies was the dog, which was prepared with a pouch of the body of the stomach which was both vagally and sympathetically denervated. The gastric remnant was drained by a large-diameter gastric cannula. A vagally denervated antral pouch was also constructed and a cannula was inserted into the fundus of the gall bladder. This model allowed the simultaneous measurement of motor activity in the pouch of the antrum of the stomach, acid and pepsin secretion and intra-gall bladder pressure changes.

The two preparations were administered in doses which gave comparable gall bladder-contracting effects of about 50% of the maximum response (Fig. 1). A significantly different effect on antral motor activity between the two preparations could not be observed.

However, the purer preparation of CCK-PZ (40% pure) produced a greater stimulatory effect on acid secretion from the denervated gastric pouches than the 10% pure material (Fig. 2). Two possible explanations for the uncoupling

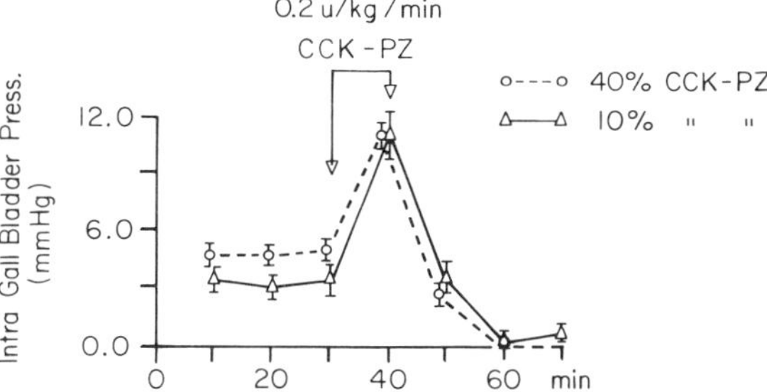

Fig. 1. Gall bladder pressure changes in the dog after infusion of either 10% or 40% pure CCK-PZ at a dose of 0.2 IDU/kg per minute. The amount of the preparations infused was equal with respect to gall bladder-contracting activity

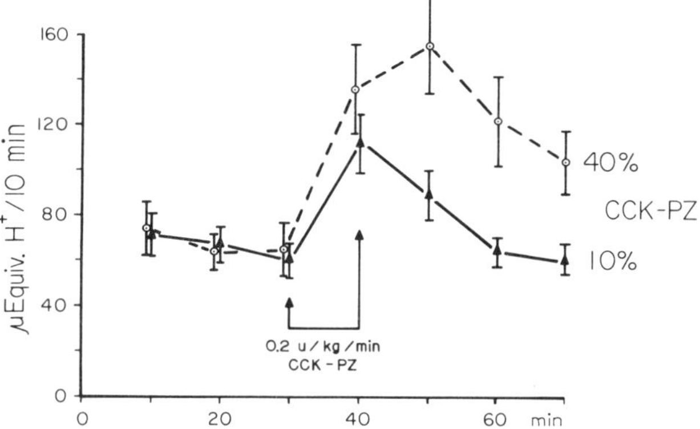

Fig. 2. Changes in acid secretion from Bickel pouches in dogs, following the intravenous infusion of 10% and 40% pure preparations of CCK-PZ. The doses were equipotent with respect to gall bladder-contracting activity but gave significantly different acid stimulatory responses

of the gall bladder and acid stimulatory effects were proposed. Either a stimulator for gastric acid secretion was selectively concentrated or an inhibitor for acid secretion was removed during the purification procedure. The former possibility was considered to be less attractive because of the structural similarities between CCK-PZ and gastrin and because both possessed similar C-terminal amino acid sequences in which resided the biologically active site for acid secretion. In addition, limited testing of a preparation of CCK-PZ with an activity of approximately 3000 IDU/mg was described as also having an acid stimulatory effect in this animal model. Two different preparations of CCK-PZ which nevertheless had approximately the same cholecystokinin activity as the preparations used earlier, i.e. 250 IDU and 1500 IDU/mg, were tested for acid inhibitory activity in the same animal model. However, in this instance acid secretion from the gastric pouch was stimulated with a dose of pentagastrin, calculated to give approximately 60% of maximum acid secretion, prior to infusion of the preparations. The same dose of each, with respect to gall bladder-contracting activity, was infused over a 10-min period. The less pure preparation of CCK-PZ (approximately 10%) was a more potent inhibitor of acid secretion than the purer preparation (approximately 40% pure)

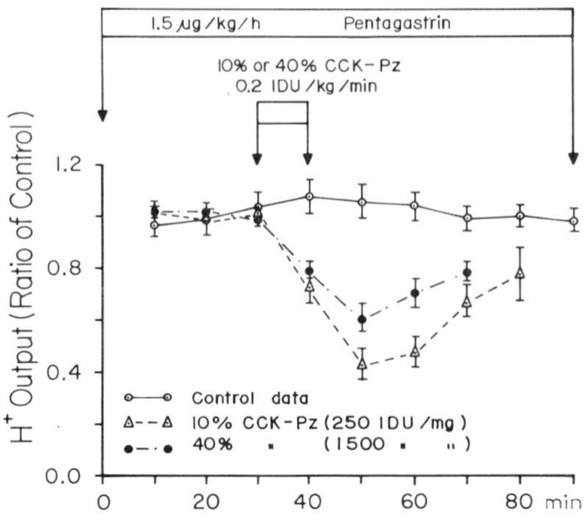

Fig. 3. Acid secretory responses from Bickel pouches, in dogs; to the intravenous infusion of 1.5 µg/kg pentagastrin per hour and the inhibitory effect of infusions of 10% and 40% CCK-PZ. Significantly greater inhibition of acid secretion was produced by the less pure material

(Fig. 3). These studies added support to the hypothesis that there was an inhibitor for acid secretion in CCK-PZ-containing preparations other than CCK-PZ itself.

2. Tissue Extraction Techniques

The hypothesis adopted by Brown and Pederson (1970), that CCK-PZ preparations contained an inhibitor for gastric acid secretion in dogs prepared with vagally and sympathetically denervated pouches of the body of the stomach, was supported by parallel studies being undertaken by Brown et al. (1969). An approximately 10% pure CCK-PZ material prepared as described by Jorpes and Mutt (1961) was subjected to chromatography on Sephadex G 50 with 0.25 *M* phosphate buffer at pH 8.0. All fractions obtained were assayed for cholecystokinin activity. Those fractions without cholecystokinin activity were pooled. Peptide material was precipitated by saturation with NaCl, and the precipitated material was rechromatographed on Sephadex G 25 with 0.2 *M* acetic acid as eluting buffer. Three fractions (I, II and III) were obtained, and fraction II was found to demonstrate little cholecystokinin activity but to produce profound inhibition of acid secretion at doses of 1.0 µg/kg. Separation of acid inhibitory and cholecystokinin activity had been achieved.

This separation procedure proved unsatisfactory for further purification attempts, and a modified approach was adopted. This technique was described by Brown et al. (1970) and resulted in the production of material of a purity suitable for qualitative amino acid analysis and bioassay. One stage of purification involved fractionation on carboxymethylcellulose (CM 11) with elution with 10 m*M* NH_4HCO_3 buffer at pH 7.8, followed by 200 m*M* NH_4HCO_3 buffer at the same pH. Three fractions were obtained, being designated EGI fractions A, B and C. These fractions were subjected to bioassay for cholecystokinin activity using the method of Ljungberg (1962) and for enterogastrone activity (Brown and Pederson 1970). Fraction B demonstrated a potent acid inhibitory effect with little cholecystokinin activity (Fig. 4), whereas

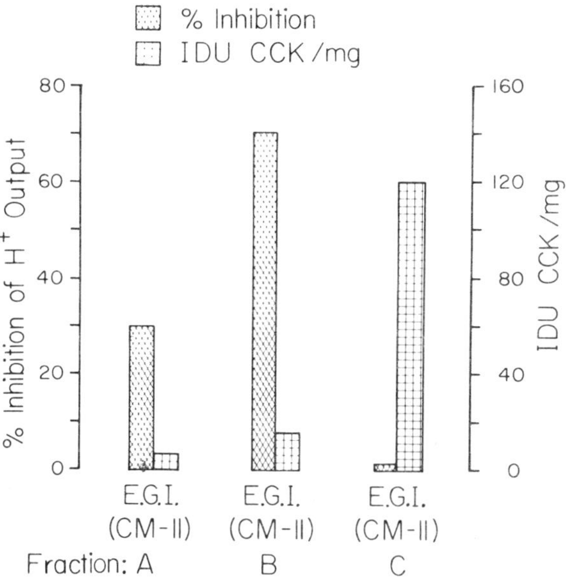

Fig. 4. Acid inhibitory and gall bladder-contracting assays of the fractions produced during ion-exchange chromatography of GIP starting material on carboxymethyl-cellulose (CM-11). An in vitro guinea-pig gall bladder was used for the assay of CCK-like activity and inhibitory activity for acid secretion was assayed in a Bickel pouch dog. CCK-like activity is expressed in Ivy Dog Units (IDU). Acid inhibitory acitivity is expressed as a percentage of the inhibition of pentagastrin-stimulated acid secretion produced by 1.0 μg/kg per hour of each fraction. E.G.I. refers to stage I material in the purification process

fraction C had little acid inhibitory activity but most gall gladder-contracting effect. The fractionation procedure had produced preparations in which the cholecystokinin-like activity had been separated from the acid inhibitory effects, confirming the hypothesis that different substances produced these effects.

B. Chemistry

I. Isolation and Purification

1. Isolation

The early attempts at purification of gastric inhibitory polypeptide (GIP) were more or less completely concerned with demonstrating that fractions with acid inhibitory activity could be separated from others demonstrating gall bladder-contracting activity.

Figure 5 summarizes the stages involved in the preparation of GIP starting material. The method is essentially that used by Jorpes and Mutt (1961) in the preparation of secretin and CCK-PZ. The first metre of the duodeno-jejunal region of the small intestine of the hog was used. Prior gentle washing through with cold tap water, followed by boiling for 5-10 min, ensured coagulation of the bulk of the protein and destruction of proteolytic enzymes

```
HOG DUODENO JEJUNAL TISSUE
↓
heat coagulation
↓
acetic acid extraction
↓
reduce pH to <2.5 with 2M HCl
↓
adsorption to alginic acid
↓
wash with 5.0 mM HCl
↓
elute with 200 mM HCl
↓
precipitate by saturation with NaCl
↓
extraction with methanol → METHANOL SOLUBLE
↓
METHANOL INSOLUBLE FRACTION
↓
CM-cellulose pH 6.5
↓
CRUDE CCK-PZ
↓
TEAE-cellulose pH 9.1
↓
10% PURE CCK-PZ
↓
Sephadex G 50
↓
GIP STARTING MATERIAL
```

Fig. 5. Summary of the stages involved in the preparation of the starting material for GIP purification

which would otherwise rapidly degrade the hormones. This procedure also served to aid the extraction with dilute acetic acid because coagulated proteins are insoluble in these conditions, thus excluding excessive amounts of inactive proteinaceous material (Mutt 1959).

Methanol extraction was carried out on material which had been precipitated at pH 4.5 by saturation with NaCl. The methanol soluble material contained secretin, vasoactive intestinal peptide (VIP) and motilin, whereas CCK-PZ and GIP were methanol insoluble under these conditions. Separation of GIP from CCK-PZ was achieved using Sephadex G50 (Brown et al. 1970). The steps involved in the purification of the starting material are shown in Fig. 6. Amino acid composition and sequence work were performed on stage IV material.

2. Purification Criteria

a) Polyacrylamide Gel Electrophoresis

The method used was a modification of that described by Johns (1967). After preparation, the gel mixture was degassed under gentle vacuum for 30 min and the 5.0 × 75 mm gel columns were poured. Distilled water was layered on top of each gel, and polymerization was accelerated under direct light for 60 min and then at room temperature. The gel tubes were capped and stored

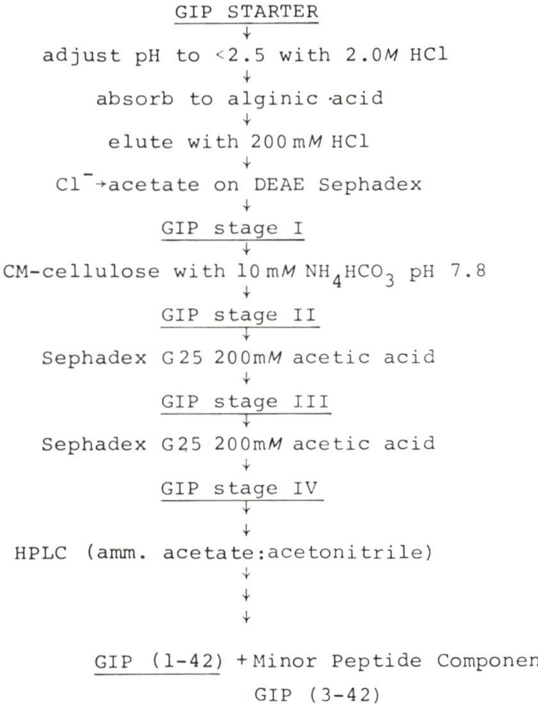

Fig. 6. Summary of the stages involved in the purification of GIP from the starting material

for at least 72 h before use. Gels were equilibrated for 3 h in 10 m*M* acetic acid at 320 V before application of the sample. Samples for electrophoresis were dissolved in 2 m*M* acetic acid with 1.0 *M* sucrose. Samples were settled into the gel by application of 320 V for 15 min. One millilitre of 0.5% w/v amido black in 1.0 *M* acetic acid was added to the lower reservoir and a voltage of 320 V applied for a further 15 min. The reservoirs were then carefully emptied and refilled with 10 m*M* acetic acid, and current was allowed to pass until the gel background cleared of dye. The peptide bands became fixed by the counter-passage of dye. Invariably this technique revealed the presence of a single peptide when applied to stage II, III or IV GIP preparations.

b) Thin Layer Chromatography

All GIP preparations were subjected to analysis by thin layer chromatography (TLC) on pre-coated silica gel plates (Mallinckrodt Chromar 7GF) initially in a solvent system of distilled water : *n*-butanol : pyridine : acetic acid (12 : 15 : 10 : 3). This technique showed GIP III and IV preparations as homogeneous materials with an R_F value of 0.83. Similar observations were made when GIP preparations were subjected to reversed-phase TLC using Whatman KC_{18} plates and a solvent system of 60% 10 m*M* ammonium acetate pH 4.0 in 500 m*M* NaCl : 40% ethanol. Single discrete spots were obtained following staining with ninhydrin/cadmium acetate reagent, with an average R_F value of 0.78. When TLC on silica gel plates was performed in a solvent system of distilled water : *n*-butanol : pyridine : acetic acid (18 : 15 : 10 : 3), i.e. increasing the water content by 50%, two discrete components could be separated. The major component had an R_F value of 0.70 and the minor, 0.66.

c) High Pressure Liquid Chromatography

High pressure liquid chromatography (HPLC) has been performed on a μBondapak C_{18} column (3.9 × 300 mm) using a Waters pump and a model 450 variable wave length detector. The solvent system developed for this system was 40 m*M* ammonium acetate, pH 4.0, and acetonitrile in a ratio of 7 : 3. In preparative runs, the elution speed was 2.0 ml/min at a pressure of 200 psi. Sample sizes varied from 5 to 200 μg and were dissolved in water to a final concentration of 2.0 μg/μl. Detection was at a wavelength of 225 nm. Two distinct peaks can be separated in this system and occasionally a third can be seen. The minor peptide component has a retention time of approximately 27 min, and the major peptide component, a retention time of approximately 37 min (Fig. 7).

d) Capillary Isotachophoresis

Capillary isotachophoresis has been performed on GIP preparations using a LKB 2127 Tachophor. The following electrolyte compositions and operating conditions have been found to be satisfactory for GIP. The leading electrolyte was 10 m*M* potassium acetate adjusted to pH 5.0 in 0.25% hydroxypropyl methylcellulose. The terminating electrolyte was 10 m*M* α-alanine. The runs were executed at 10 °C at a constant amperage of 50 μA and detection was by u.v. light absorbance at 280 nm. Two peptide components could be iden-

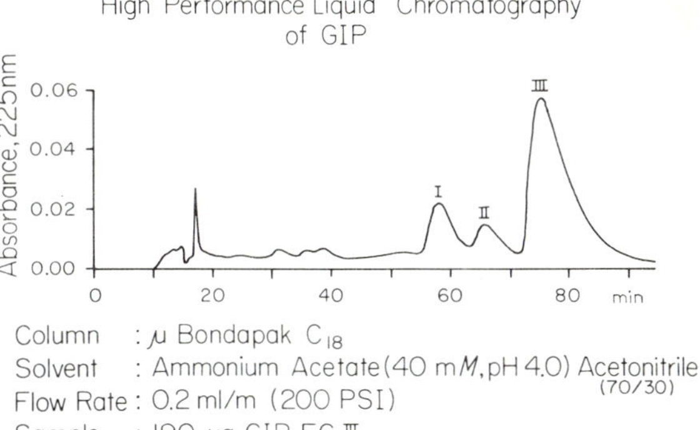

Fig. 7. High performance liquid chromatography (HPLC) profile of 190 μg GIP(III) on a μBondapak C_{18} column. Absorbance was read at 225 nm. Peak III corresponds to the peptide with the amino acid sequence of $GIP_{(1-42)}$, and peak I has the amino acid sequence $GIP_{(3-42)}$

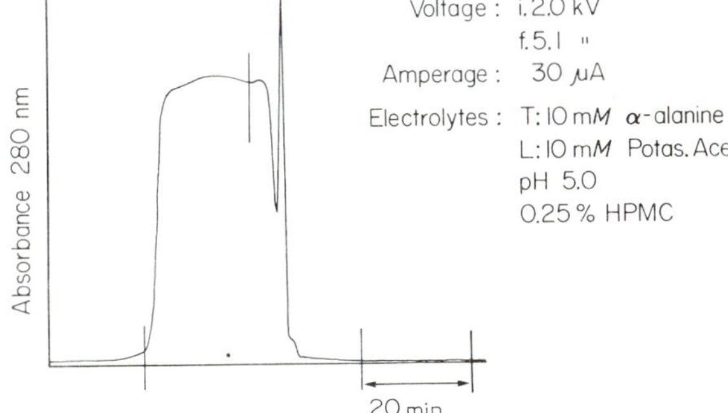

Fig. 8. Isotachophoresis of 40 μg (GIP (III). Absorbance read at 280 nm. The initial voltage was 2.0 kV, rising to 5.1 kV at a constant amperage of 30 μA. *HPMC*, hydroxypropyl methylcellulose

tified; the major component accounted for about 95 % of the total absorbance and was less mobile (Fig. 8).

The application of more sophisticated techniques for the isolation and qualitative assessment of peptides to GIP preparations revealed that there was consistently a minor peptide component present which could not be removed by conventional ion exchange chromatography and gel filtration.

II. Amino Acid Sequence

The amino acid sequence of GIP was determined using a preparation of GIP (stage IV) which had been shown to be homogeneous by polyacrylamide gel electrophoresis and high-voltage electrophoresis at pH 6.5.

1. Cyanogen Bromide Cleavage

Preliminary amino acid analyses had revealed the presence of a single methionine residue. Cyanogen bromide (CNBr) cleavage of the molecule at this methionine residue (Brown 1971) resulted in the production of two peptides of chain lengths of 14 and 29 amino acids. A two-stage system using Sephadex G 25 fine to separate the smaller peptide from the reaction mixture and carboxymethyl cellulose CM-11 to separate the larger peptide from uncleaved GIP resulted in products of high purity, suitable for amino acid analyses and sequence work.

The smaller CNBr-cleaved peptide proved to be the N-terminal fragment of the molecule, with tyrosine as the N-terminal residue. The amino acid sequence was determined using the dansyl-Edman degradation technique on the whole peptide, and confirmation was obtained by sequence analysis of chymotryptic peptides. Brown and Dryburgh (1971) reported this sequence to be:

NH_2–Tyr–Ala–Glu–Gly–Thr–Phe–Ile–Ser–Asp–Tyr–Ser–Ile–Ala–Met.

The C-terminal CNBr fragment proved to be too long to sequence by the Edman technique. The N-terminal residue of this fragment was determined to be Asp/Asn and eventually shown to be Asp. The following amino acid composition was arrived at:

Ala(1), Arg(1), Asx(5), Glx(5), Gly(1), His(1), Ile(2), Leu(2), Lys(5), Phe(1), Ser(1), Thr(1), Trp(2), Val(1).

2. The Tryptic Peptides

Elucidation of the greater part of the amino acid sequence of GIP was arrived at by sequence studies with peptides produced by tryptic digestion of both the complete molecule and the C-terminal CNBr fragment. Brown (1971) described the methods whereby the peptides produced by tryptic digestion could be separated and purified. These techniques included cation and anion exchange chromatography and countercurrent distribution. The amino acid compositions and N-terminal residues were reported, and the complete amino acid sequences published later (Brown and Dryburgh 1971). Table 1 summarizes

Table 1. The amino acid sequences of the tryptic peptides of porcine GIP

Peptide		Rf[a] TLC	Electrophoretic mobility[b]
Tr2	NH_2 Tyr-Ala-Glu-Gly-Thr-Phe-Ile-Ser-Asp-Tyr-Ser-Ile-Ala-Met-Asp-Lys	0.66	−0.53
Tr3a	NH_2 Ser-Asp-Trp-Lys-His-Asn-Ile-Thr-Gln	0.56	0.03
Tr3b	NH_2 Gln-Gln-Asp-Phe-Val-Asn-Trp-Leu-Leu-Ala-Gln-Gln-Lys	0.71	0.03
Tr3c[c]	NH_2 His-Asn-Ile-Thr-Gln		
Tr4[c]	NH_2 Lys-Ser-Asp-Trp-Lys	0.47	0.35
Tr5	NH_2 Il-Arg	0.62	0.54
Tr6[c]	NH_2 Gly-Lys	0.44	0.83
Tr7	NH_2 Gly-Lys-Lys	0.39	0.94

[a] Solvent system used for TLC was distilled water: *n*-butanol:pyridine:acetic acid (12:15:10:3)
[b] Electrophoretic mobilities at pH 6.5 with respect to aspartic acid distance travelled by peptide ∕ distance travelled by aspartic acid
[c] Minor tryptic cleavages

the sequences of the tryptic peptides and includes the electrophoretic mobilities at pH 6.5 and R_F values from TLC on silica gel. Sequence confirmations were obtained from peptides produced by chymotrypsin digestion of the whole molecule or of the C-terminal CNBr fragment. In particular the tryptophan-containing peptides were isolated, following detection by Ehrlich's reagent, by paper chromatographic techniques and sequenced using the dansyl-Edman procedure.

Alignment of the tryptic peptides was rendered relatively easy because of the number of minor cleavages which were produced following 6 h hydrolysis with trypsin (Brown 1971). The amino acid sequence arrived at from these studies was reported to be:

NH_2–Tyr–Ala–Glu–Gly–Thr–Phe–Ile–Ser–Asp–Tyr–Ser–Ile–
Ala–Met–Asp–Lys–Ile–Arg–Gln–Gln–Asp–Phe–Val–
–Asn–Trp–Leu–Leu–Ala–Gln–Gln–Lys–Gly–Lys–Lys–Ser–
Asp–Trp–Lys–His–Asn–Ile–Thr–Gln

III. Correction to Sequence

Moroder et al. (1978) suggested, as one possibility, that an error in the published sequence could be responsible for the poor acid inhibitory effect of GIP (1-38), which had powerful insulinotropic activity. Another possible explanation for the results otained with GIP (1-38) was that the purest GIP preparations could in fact contain two closely related polypeptides, one with inhibitory activity for gastric acid secretion and the other with insulinotropic activity. Brown et al. (1981) demonstrated with HPLC and isotachophoresis that GIP preparations earlier considered to be homogeneous on the basis of conventional techniques did in fact contain a minor peptide component. For these reasons, Jörnvall et al. (1981) reinvestigated the primary structure of a preparation of natural procine GIP. The previously suggested sequence was shown by direct liquid-phase sequencer analysis to be largely correct, except for the suggestion that there should be one instead of two glutamine residues at position 29-30. Confirmation of this change was obtained by liquid-phase sequencer degradation of the fragments produced by cleavage of GIP with CNBr. The degradation was performed without separation of the peptides. Two major sequences were determined, as expected from the single methionine residue. Confirmatory sequences were obtained by manual dansyl-Edman analyses on a tryptic digest of GIP, again without separation of the peptides, and by cleavage at tryptophan residues with *N*-chlorosuccinimide followed by separation of the products on HPLC. The presence of only one glutamine residue at position 29-30 was confirmed. The corrected sequence was reported as stated below (see next page) and it was suggested that this was the sequence of the major component in GIP preparations, the fraction III from HPLC referred to by Brown et al. (1981). Sufficient material of HPLC fraction III was obtained to bioassay for insulinotropic activity in the isolated perfused rat pancreas. This fraction was found to posses insulinotropic activity of a similar magnitude to GIP when infused at a concentration of

1 5 10
Tyr–Ala–Glu–Gly–Thr–Phe–Ile–Ser–Asp–Tyr–Ser–Ile–
15 20
–Ala–Met–Asp–Lys–Ile–Arg–Gln–Gln–Asp–Phe–Val–Asn–
25 30 35
–Trp–Leu–Leu–Ala–Gln–Lys–Gly–Lys–Lys–Ser–Asp–Trp–
40
–Lys–His–Asn–Ile–Thr–Gln

5.0 ng/ml. Insulinotropic activity was not found in the HPLC fraction I (Brown et al. 1981).

The sequence analyses also showed that the GIP preparation was heterogeneous, with a much higher concentration (20%) of a minor component than reported by Brown et al. (1981). The sequence analyses suggested that the amino acid sequence of the minor component was identical to residues 3-42 of the main component. The extensive similarity between these two components accounts for the difficulty in separating them.

IV. Synthesis

Gastric inhibitory polypeptide, like most of the gastrointestinal polypeptides, occurs in cells which have a diffuse distribution in the mucosa of the gastrointestinal tract. Yields from extraction are low, and the polypeptide cannot be made readily available. A successful synthesis would serve to increase the availability of the peptide and also confirm the published sequence, but to date such a synthesis has not been satisfactorily completed.

Camble et al. (1973) and Yajima et al. (1975) attempted a total synthesis of the tritetracontapeptide corresponding to the published sequence. Little assessment of these products was made. Yanaihara et al. (1978) produced a synthetic preparation of GIP which they subjected to purification procedures on DEAE-cellulose with ammonium acetate buffer and DEAE-Sephadex A25 with tris-glycine buffer. The partially purified synthetic tritetracontapeptide showed approximately 30% insulinotropic activity as natural GIP in the isolated perfused rat pancreas preparation, significant inhibition of pentagastrin-stimulated acid secretion in the dog and approximately 30% immunoreactivity. After DEAE-Sephadex A25 chromatography they were able to show an increase in immunoreactivity to 52% in comparison with natural porcine GIP. When this further purified product was digested with trypsin and subjected to high-voltage electrophoresis, the electrophoretic mobilities were found to be reasonably comparable to those of natural porcine GIP.

Moroder et al. (1978) described the synthesis of an octatriacontapeptide corresponding to the sequence 1-38 of porcine GIP. The purified product demonstrated insulinotropic activity in the isolated perfused rat pancreas, identical to that of the natural porcine hormone, but had a poor acid inhibitory effect in the dog. Moroder et al. cited three possible explanations for the lack of acid inhibitory activity. The reduced gastric effect could have been due to the lack of the C-terminal tetrapeptide or an error in the published sequence of the natural material. The third possibility was that the natural porcine GIP preparation could in fact contain a second component which has the gastric inhibitory activity.

Confirmation of the sequence of natural porcine GIP by production of a synthetic analogue with full biological activities has not, as yet, been completed.

C. Physiological Actions of Exogenous GIP

I. Gastrointestinal Effects

1. Gastric Inhibitory Activity

a) Inhibition of Acid Secretion

In the early studies on the purification of GIP, the parameter measured was the ability of the extracts to inhibit acid secretion. The animal model which was used throughout these studies was the vagally and sympathetically denervated pouch of the body of the stomach of the dog (Brown and Pederson 1970). The stimulus for acid secretion was pentagastrin administered at a dose calculated to produce a plateau of acid output which was 60%-70% of the maximum for this stimulus (Pederson and Brown 1972). Figure 9 shows the inhibitory effect on a plateau of acid secretion when three defferent doses of GIP (0.25, 0.5 and 1.0 μg/kg) were infused intravenously over a 1-h period. When the stimulus for acid secretion in this model was histamine dihydrochloride, infusion of GIP would still induce inhibition of acid secretion, but the effect was less marked and higher doses of GIP had to be employed (4.0 μg/kg over 1 h). In the vagally innervated gastric remnant, acid secretion induced by insulin hypoglycaemia could be reduced by the infusion of 4.0 μg/kg GIP over 1 h. The inhibitory effect was, again, much weaker than against pentagastrin, but similar to that observed with histamine. Soon-Shiong et al. (1979a) observed that in the innervated stomach preparation of dogs GIP demonstrated only a

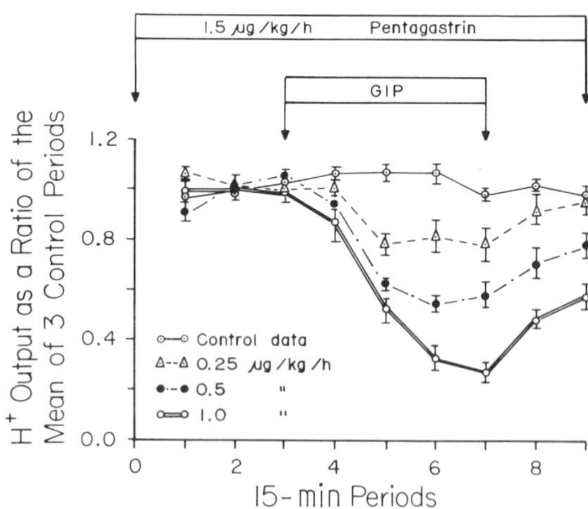

Fig. 9. The effect of a 60 min infusion of GIP in doses of 0.25, 0.50 and 1.0 μg/kg on acid secretion from Bickel pouches of dogs. Acid secretion was stimulated by pentagastrin administered at a dose of 1.5 μg/kg per hour for the duration of the experiment and is expressed as a ratio of the mean of three control periods

weak inhibitory activity against pentagastrin-stimulated acid secretion. Maxwell et al. (1980) also observed only weak inhibition of pentagastrin-stimulated acid and pepsin secretion in man when innervation was intact. Soon-Shiong et al. (1979b) were able to reproduce the profound inhibition of pentagastrin-stimulated acid secretion in vagally denervated gastric pouches of the dog, as had been earlier described by Pederson and Brown (1972), and showed that intravenous infusion of 25 μg/kg urecholine per hour could abolish the inhibitory action of GIP. These studies suggested to Brown et al. (1980) the possibility that GIP acts indirectly on the parietal cell, via a mechanism which can be regulated by parasympathetic innervation. This hypothesis has received considerable support recently. Radioimmunoassay and immunohistological studies have demonstrated that the stomach contains high concentrations of somatostatin-like immunoreactivity (SLI) (McIntosh and Arnold 1978; Efendic et al. 1978). Further, it has been shown that exogenous somatostatin is a potent inhibitor of acid secretion from the stomach. McIntosh et al. (1979) have developed an isolated perfused rat stomach preparation and used it to study the possibility that GIP exerted its acid inhibitory action via the local release of SLI. Figure 10 shows that the preparation spontaneously released SLI at a gradually declining rate and that infusion of porcine GIP at concentrations of 5 and 50 ng/ml produced dose-related increases in SLI secretion.

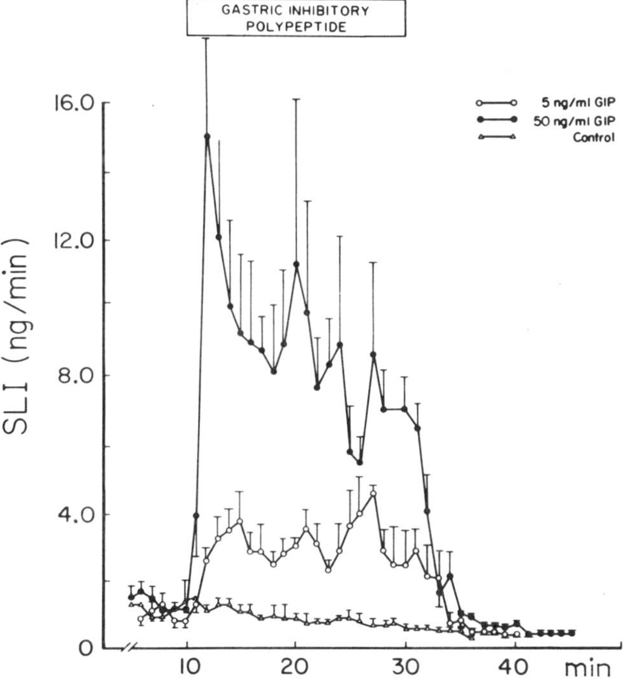

Fig. 10. The effect of porcine GIP on somatostatin-like immunoreactivity (SLI) from the isolated perfused stomach of the rat. The stomach was perfused with a Krebs-Ringer bicarbonate buffer with 0.2% HSA and 3.0% dextran T70

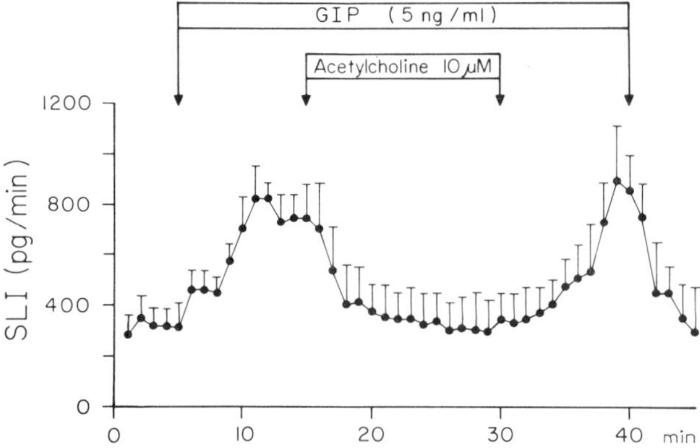

Fig. 11. The effect of acetylcholine on the release of SLI by GIP in the isolated perfused rat stomach preparation. A prompt inhibition of SLI release was produced by the acetylcholine and was maintained for the duration of the infusion period

The demonstration by Soon-Shiong et al. (1979b) that the acid inhibitory action of GIP could be blocked by administration of a cholinomimetic substance prompted an investigation into the interelationship of GIP and vagal innervation with SLI release in the perfused rat stomach preparation. Perfusion of the rat stomach with GIP will release SLI and a plateau of SLI secretion can be maintained. When acetylcholine was introduced into the perfusate to a concentration of $10^{-7} M$ after establishing SLI release with GIP, there was a prompt and complete inhibition of SLI release (Fig. 11). A similar effect was observed when vagal stimulation (7 V, 10 Hz, 5 ms) was induced after establishment of SLI release with 5 ng/ml GIP (Fig. 12).

In the presence of intact vagal innervation, GIP has been shown to have no inhibitory effect on pentagastrin-stimulated acid secretion. The absence of this inhibitory effect occurred along with the suppression of GIP-induced SLI release by vagal stimulation and acetylcholine perfusion. GIP and vagal excitation exert reciprocal effects on SLI release, and it is highly probable that GIP exerts its inhibitory effect on the parietal cell via SLI.

b) Inhibition of Gastrin Release

In addition to the inhibitory effect of GIP on the parietal cell, an acid inhibitory function for the hormone via the inhibition of gastrin release has also been described. Villar et al. (1976) fed a standard meat meal to dogs which had been prepared with Heidenhain pouches. Radioimmunoassay measurement of gastrin showed that intravenous infusion of GIP (2 μg/kg per hour) for 2 h suppressed gastrin release in response to the meal stimulus when food was taken 60 min after the commencement of the infusion. Serum IR-GIP levels measured prior to feeding were approximately 1.8 ng/ml and became further elevated after food ingestion (to in excess of 3.5 ng/ml).

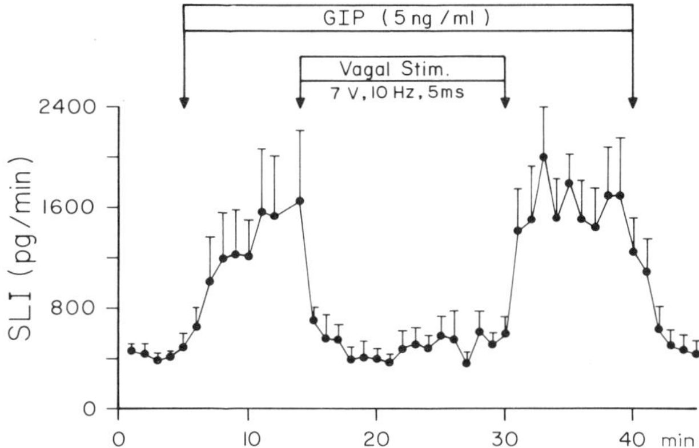

Fig. 12. The effect of preganglionic vagal stimulation on SLI release by GIP. Stimulation of the vagus nerve (7.0 V, 10 Hz, 5.0 ms) at the level of the lower oesophagus resulted in an immediate inhibition of SLI release, which was stimulated by GIP. The inhibition was maintained for the duration of the stimulation

2. Effect on Pepsin Secretion

Pederson and Brown (1972) studied the effect of different doses of GIP on pepsin output from both the vagally denervated gastric pouch and the vagally innervated gastric remnant of the dog. When pentagastrin was the stimulus, pepsin secretion from the vagally denervated pouch preparation was inhibited by the hormone in a dose-dependent manner. A larger dose of GIP, 4.0 μg/kg per hour for 1 h also inhibited pepsin secretion when the pouches were being stimulated by intravenous infusion of histamine dihydrochloride (10 μg/kg per hour).

In the vagally innervated gastric remnant, induction of hypoglycaemia by the single rapid intravenous injection of 0.5 units/kg insulin induced a large increase in pepsin secretion. When animals were infused with 4.0 μg/kg GIP per hour for 1 h commencing at the same time as the insulin injection and for 60 min after injection, pepsin output was significantly reduced. Maxwell et al. (1980) were unable to demonstrate an inhibitory action of GIP on pepsin secretion in man when varying doses of pentagastrin were administered for 30-min periods. The dose of GIP administered was 2.0 μg/kg per hour. The plasma concentration of GIP, measured by radioimmunoassay, showed a mean ± SE plateau level of 7.4 ± 1.4 ng/ml, which is much higher than those levels observed following a meal. It is apparent that the role of GIP in the regulation of pepsin secretion has striking similarities to that observed in regulation of acid secretion, in that little effect can be observed when the experimental model has vagal innervation intact.

3. Intestinal Secretion

Nasset et al. (1935) demonstrated that the presence of food substances in the upper small intestine stimulated secretion from isolated denervated intestinal loops. A humoral mechanism was postulated and named enterocrinin (Nasset

1938). Wright et al. (1979) confirmed the existence of such a humoral mechanism using balloon exclusion experiments.

Barbezat and Grossman (1971) intravenously infused GIP at a dose of 9.0 μg/min for 15 min into dogs with 30-min Thiry-Vella fistulae of the upper jejunum or lower ileum. Secretion rates from both jejunal and ileal fistulae were increased. It is now known that this dose of GIP would have produced supra-physiological levels of circulating peptide.

Helmon and Barbezat (1977) used a triple lumen gut perfusion technique to investigate the effect of GIP on human jejunal, water and electrolyte transport. The dose of GIP which was infused intravenously was 1.0 μg/min for 30 min. This dose had been shown by Dupré et al. (1973) to elevate serum IR–GIP levels to within the range achievable in normal subjects following ingestion of a meal (Kuzio et al. 1974). A net reduction of water absorption from 62.9 $\pm$6.5 to 18.5$\pm$4.2 μl/min per centimetre resulted. This reduction in water absorption was associated with a change in chloride flux, which was reversed from absorption of 4.36$\pm$0.25 μ*M*/min per centimetre during the control periods, to secretion of 1.56$\pm$0.65 μ*M*/min per centimetre. Na^+, K^+ and HCO_3^- absorption were significantly reduced during the infusion of GIP, but returned to pre-infusion levels when GIP administration was discontinued. Helman and Barbezat (1977) concluded that the type of response observed after GIP infusion was unlike that seen with cyclic adenosine monophosphate (cAMP)-mediated cholera toxin-induced secretion as described by Carpenter et al. (1968). The recognition of a difference in the response to GIP and cholera toxin was supported by the observation of Schwartz et al. (1974), who were unable to show an elevation of adenylate cyclase activity in the gastro-intestinal tract in response to GIP and suggested that its mode of action must be via an alternative pathway.

4. Salivary Secretion

Denniss and Young (1978) studied the effects of polypeptides on electrolyte transport in the main excretory duct of the mandibular gland of the rabbit. They observed that GIP at a concentration of 10^{-11} mol/litre reduced the net movement of Na^+ from lumen to interstitium by 18.5% and the net Cl^- movement by 10% but failed to alter the transepithelial potential difference even at a concentration of 10^{-7} mol/litre. No effects were observed on fluxes of K^+ or HCO_2^-.

Denniss and Young postulated that because the effects were seen at physiological concentrations of GIP, it could possibly play a role along with VIP and other gut hormones in determining the electrolyte composition of saliva.

5. Mesenteric Blood Flow

In the anaesthetized cat, the intravenous infusion of GIP produced a dose-dependent increase in superior mesenteric blood flow (Fara and Salazar 1978). In this preparation, superior mesenteric blood flow at rest averaged 18.9 ml/min per kilogram body wt. and became elevated within 1 min of starting intravenous GIP infusion. It remained elevated during infusion periods of up to 20 min, but returned gradually to control after cessation of infusion. threshold dose for GIP was 250 ng/min. Arterial pressure, measured from th femoral artery, remained essentially unchanged, thus indicating a fall in mesenteric vascular resistance.

Fara and Salazar concluded that it was possible that GIP exerted a local regulatory action on the mesenteric circulation and could actually be responsible for the vasodilator effects earlier ascribed to CCK-PZ (Fara et al. 1972).

II. Metabolic Effects

1. Insulin Release

Brown and Otte (1978, 1979a, b) have recently reviewed the equivocal nature of the evidence which had suggested an insulinotropic role for gastrin, secretin and CCK-PZ. They drew particular attention to the lack of information concerning the purity of the hormone preparations used, the magnitude of the dose, the method by which it was administered and the prevailing serum glucose concentrations. Creutzfeld (1979) reviewed the literature on the incretin concept and concluded that the search for incretin was a search for an endocrine produced in the gastrointestinal tract and released by nutrients, especially carbohydrates, which stimulated insulin secretion in the hyperglycaemic state.

Rabinovitch and Dupré (1972) were able to compare the insulinotropic and glucagonotropic activities of highly purified CCK-PZ with a preparation containing 10% of the cholecystokinin activity in an in vitro perifused rat pancreas preparation and normal human volunteers. In both preparations the highly purified cholecystokinin was unable to elicit either insulin or glucagon secretion, whereas both effects were observed with the 10% pure material. These observations were similar to those described by Brown and Pederson (1970) for acid secretory effects and prompted Dupré et al. (1973) to investigate an insulinotropic action for GIP.

Figure 13 shows the mean changes in plasma glucose and IRI in normal human volunteers receiving an intravenous infusion of 0.5 g/min glucose for 60 min, with and without an intravenous infusion of 1.0 μg/min highly purified porcine GIP. This was the first series of experiments in which purified GIP had been administered to human volunteers. No subjective effects were observed by the volunteers. This dose of GIP produced a significant improvement in the glucose tolerance curve and a marked increase in IRI. In fasted normoglycaemic subjects, this dose of GIP produced only a small and transient increase in IRI release, indicating the glucose-dependent nature of this response. The above-stated dose of GIP was of a magnitude to achieve serum IR–GIP levels of approximately 1.0 ng/ml. When GIP was infused to achieve serum levels which were supraphysiological, transient insulinotropic effects in dog (Pederson et al. 1975b) and rat (Pederson and Brown 1976) were obtained. The glucose-dependent nature of the insulinotropic action of GIP can be seen in Fig. 14. At a constant perfusion level of GIP, increasing the glucose concentra-

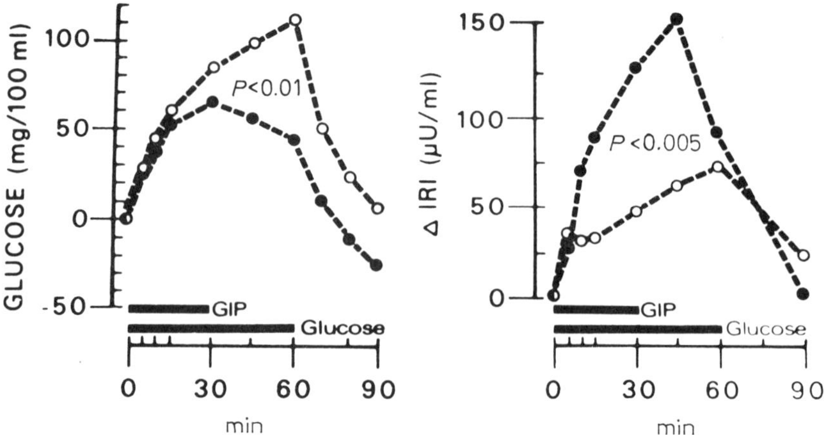

Fig. 13. A comparison of the insulinotropic effect of glucose in the presence (●) and absence (○) of GIP (1.0 μg/min), in paired studies in six normal human volunteers

tion invoked a much greater insulin release. The threshold glucose concentration in rat was shown to be approximately 5.5 m*M*, which compared favourably with that described for man, i.e. 25 mg/dl above basal, by Elahi et al. (1979). The potentiating action of GIP on insulin release in the isolated perfused rat pancreas was maximal at a glucose concentration of approximately 16 m*M*. The insulin output, however, was several-fold greater than could be produced by glucose alone. At a fixed glucose concentration of 8.9 m*M*, increasing the

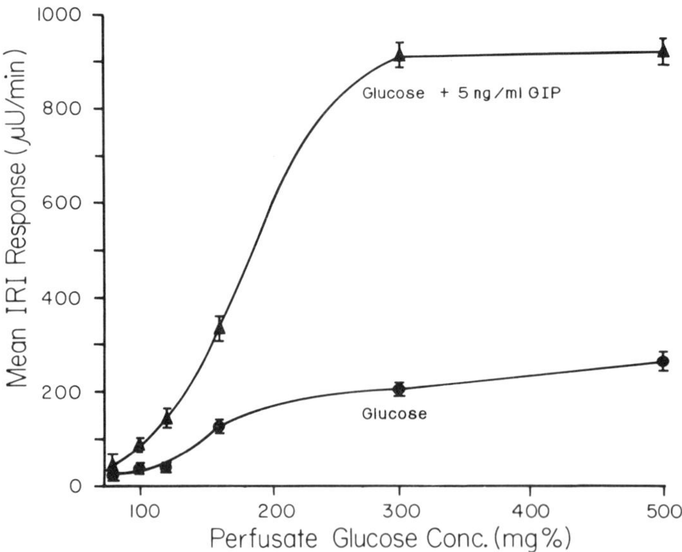

Fig. 14. The integrated IRI responses from the isolated perfused rat pancreas to graded increases in glucose concentration alone and in the presence of 5.0 ng/ml porcine GIP in the perfusate

GIP concentration of the perfusate produced a dose-dependent increase in insulin output.

Arginine will potentiate insulin release in the presence of hyperglycaemia; as in the case of GIP, the insulinotropic effect is glucose concentration dependent. Pederson and Brown (1978) showed that in the presence of 20 m*M* arginine and midrange glucose concentrations (6 m*M* and 10 m*M*) GIP still exerted an insulinotropic effect, but at concentrations less than 5 m*M* or greater than 16 m*M* there was no further increase. Pancreatic islets from the rat have been incubated in a static system with varying concentrations of glucose in the presence or absence of 10 μg/ml GIP. The glucose concentrations varied from 2.0 to 25.0 m*M* (Schauder et al. 1975). At 6.0 m*M* glucose and lower, a potentiating action of GIP was not present. Potentiation was apparent at 8.0 and 16.6 m*M* gluçose concentrations, but not at 25.0 m*M*. Brown et al. (1980) offered several possible explanations for the differences seen between the isolated perfused pancreas and isolated pancreatic islets. They suggested (a) that in the isolation of the islets the receptors for GIP may have been destroyed by the collagenase treatment; (b) that the use of a static system may have allowed the accumulation of substances inhibitory for insulin release, eg. somatostatin; (c) that absence of the intra-islet microcirculation as a means of transport in the islets; may have led to the differences; and (d) that the action of GIP on insulin release may be an indirect one.

Ipp et al. (1977) demonstrated that a GIP concentration of 58 ng/ml in the presence of 100 mg/dl glucose would stimulate somatostatin secretion from the isolated perfused dog pancreas. This dose of GIP was of a magnitude capable of stimulating insulin release in the absence of hyperglycaemia and was severalfold greater than a physiological dose. Figure 15 shows that the insulinotropic action of 1.0 ng/ml GIP on the isolated perfused rat pancreas was biphasic. There was apparently a very rapid release of insulin when GIP was introduced into the system, followed by a transitory inhibition which was comparable in

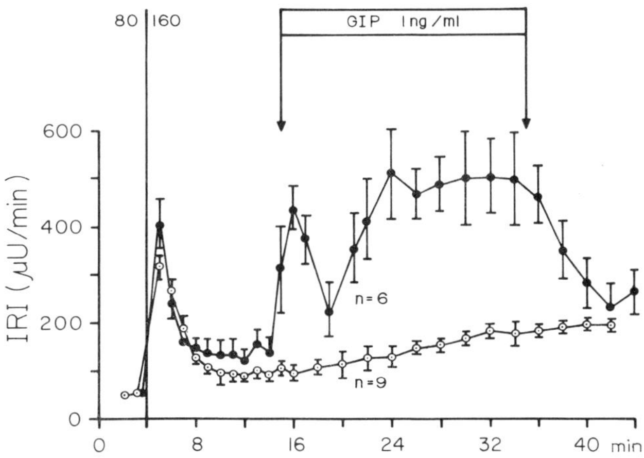

Fig. 15. The release of IRI from the isolate perfused rat pancreas in response to a perfusate glucose concentration of 160 mg/dl, with and without 1.0 ng/ml GIP. The biphasic nature of the insulinotropic action of GIP can be seen in this figure

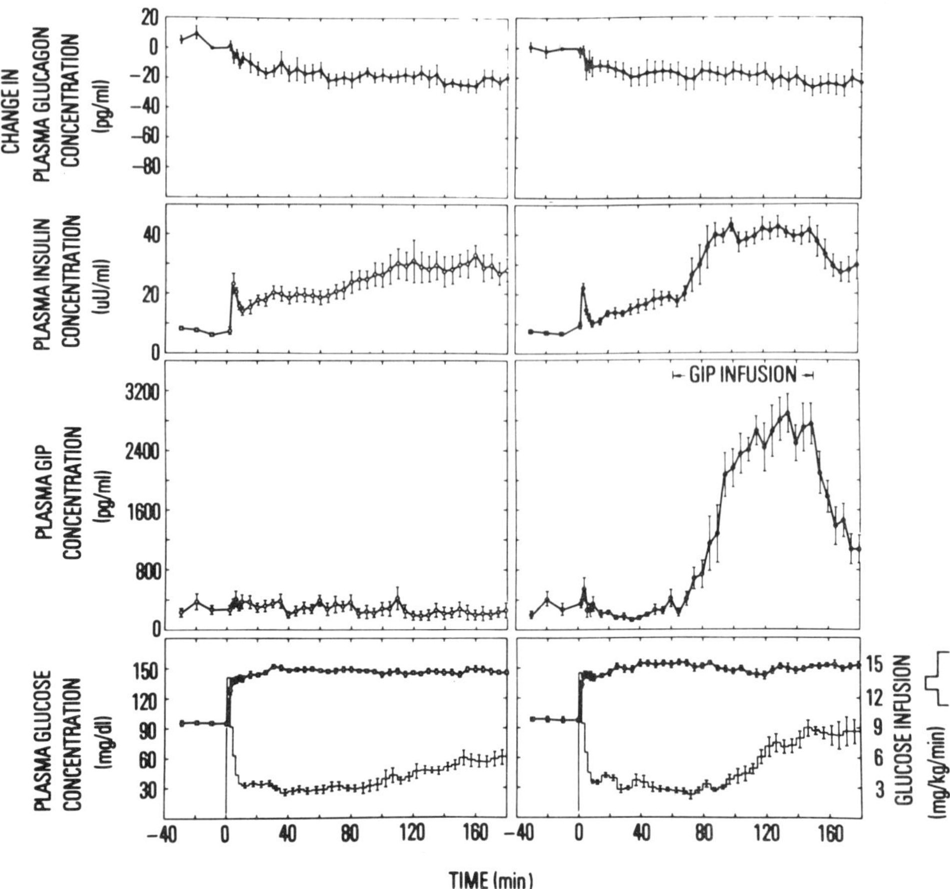

Fig. 16. Mild hyperglycaemic clamp studies in man (basal + 54 mg/dl glucose) with and without infusion of porcine GIP (6.67 ng/kg per minute). Plasma levels of glucose, IR-GIP, IRI and IR-glucagon and the amount of glucose required to maintain the stable hyperglycaemia are shown on the *left* for the group without GIP and on the *right* for the group with the GIP infusion. GIP was infused between times 60 and 150 min

time course to the first phase of insulin release. Brown et al. (1980) suggested that this 'off' phenomenon could occur as a result of the GIP being presented to the isolated perfused organ as a square wave stimulus. Under these circumstances, the pancreas would be presented with a sudden large increase in GIP, a non-physiological stimulus, which may result in a spike release of somatostatin and thereby produce a transient inhibition in insulin release.

Highly purified porcine GIP has been administered intravenously at a dose of 6.7 ng/kg per minute for 90 min in normal human volunteers (Elahi et al. 1979). Steady state conditions of hyperglycaemia were established using the glucose-clamp technique, which allowed the predetermined hyperglycaemic level to be maintained despite increasing insulin levels. GIP was administered

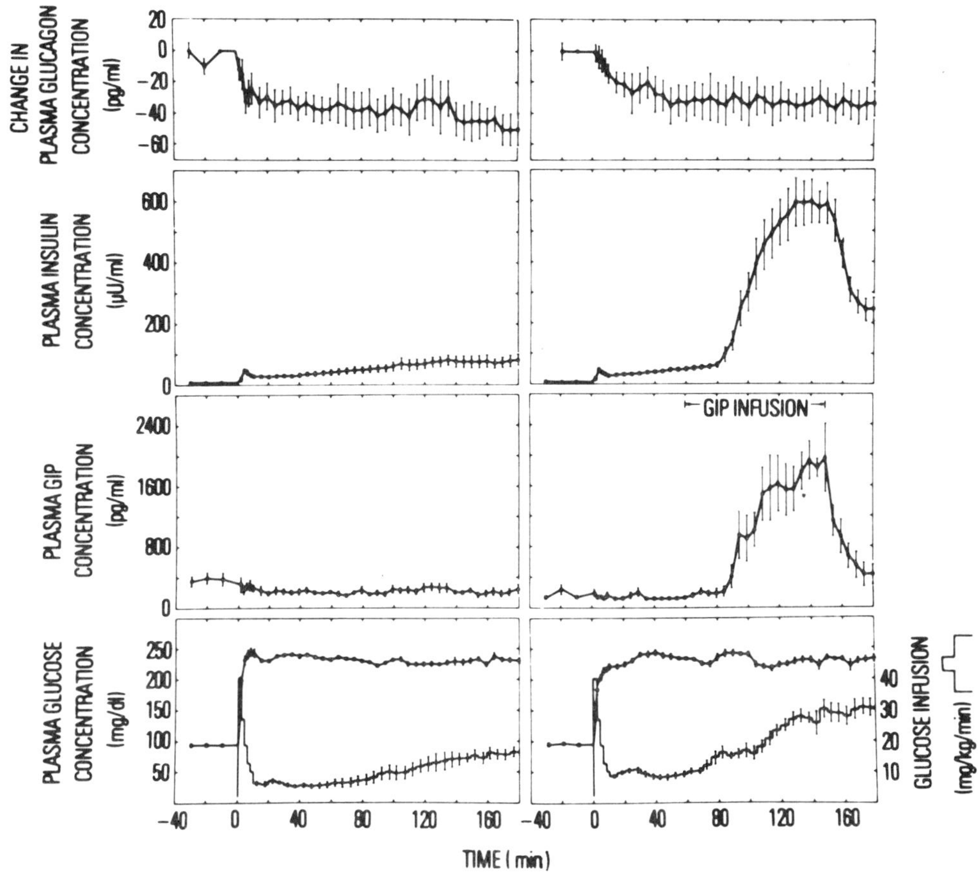

Fig. 17. Moderate hyperglycaemic clamp studies in man (basal + 143 mg/dl glucose) with and without the infusion of porcine GIP (6.67 ng/kg per minute) are shown. Plasma levels of glucose, IR-GIP, IRI and IR-glucagon and the amount of glucose infused to maintain the stable hyperglycaemia are shown on the *left* for the group without GIP and on the *right* for the group which received the GIP infusion. GIP was infused between times 60 and 150 min

during basal-state euglycaemia, after the establishment of mild, steady-state hyperglycaemia at a plasma glucose level 54 mg/dl above basal and after the establishment of moderate steady-state hyperglycaemia at a plasma glucose level of 143 mg/dl above basal. No insulinotropic effect of GIP was observed in the euglycaemic condition. In the glucose clamp experiments with mild hyperglycaemia, IRI release was elevated (Fig. 16), and at moderate hyperglycaemia, plasma IRI levels in excess of 500 μU/ml were achieved (Fig. 17). When GIP was administered it was necessary to increase the rate of glucose infusion to approximately 30 mg/kg per minute to maintain the steady-state hyperglycaemia of basal + 143 mg/dl. In the control subjects, without GIP infusion, the maximum glucose infusion rate required was 15 mg/kg per minute.

2. Glucagon Release

Unger et al. (1967) demonstrated that IR-glucagon could be released from the dog pancreas by infusion of 10% pure CCK-PZ. A potentiating effect of 10% CCK-PZ on arginine-stimulated glucagon-like immunoreactivity (GLI) has been described by Dupré et al. (1969) in man, and this effect has been shown to be suppressible by hyperglycaemia. In dog (Bottger et al. 1972) but not in man (Bottger et al. 1973) ingestion of triglyceride has been shown to stimulate glucagon secretion, an effect due to release of an intestinal humoral factor.

Brown et al. (1975b) reported that high levels of GIP (100 ng/ml) would stimulate glucagon release from the in vivo and in vitro rat pancreas. It was suggested, however, that the glucagonotropic action of GIP was not suppressible by glucose. Pederson and Brown (1978) investigated glucagon and insulin responses of the perfused rat pancreas to both glucose and arginine in the presence and absence of GIP. A glucagonotropic effect of GIP was observed at low glucose concentrations (Fig. 18). This response was suppressed by increasing the glucose concentration. The potentiation of glucagon release by GIP was not significant at glucose concentrations above 5.5 m*M*, at which the insulinotropic effect became apparent. The glucagon response to GIP in the isolated perfused pancreas in the presence of 5.5 m*M* glucose and 20 m*M* arginine was biphasic (Fig. 19), but the second phase was transitory. In man, Elahi et al. (1979) were unable to demonstrate a glucagonotropic effect of GIP when the peptide was infused at a concentration which achieved a plasma IR–GIP level in excess of 2.0 ng/ml and plasma glucose concentrations were basal, i.e. less than 100 mg/dl. In all studies with GIP, the plasma glucagon was measured using the same antibody (30 K, from R. H. Unger, Dallas, Texas).

3. Other Metabolic Effects

a) Glucose Uptake

Portal hyperinsulinaemia and hyperglycaemia both suppress glucose production by the liver. Other factors also contribute to this metabolic process. Andersen et al. (1980) studied the role of GIP in this process in dogs with chronic portal venous catheters. GIP infusion (10 ng/kg per minute) reduced hepatic glucose output without concomitant increases in serum IRI, glucose or glucose disposal levels. They suggested that GIP possessed a significant suppressive effect on hepatic glucose output in both the basal state and when it was already lowered by the effects of hyperinsulinaemia. It also potentiated the effect of insulin on net glucose disposal, and Andersen et al. concluded the GIP may act synergistically to augment the hepatic glucose response as well as the beta cell response to ingested glucose.

b) Effects on Adipocytes

In fat cells prepared by collagenase digestion of epididymal fat pads, Dupré et al. (1976) studied possible interactions between GIP and glucagon. GIP was

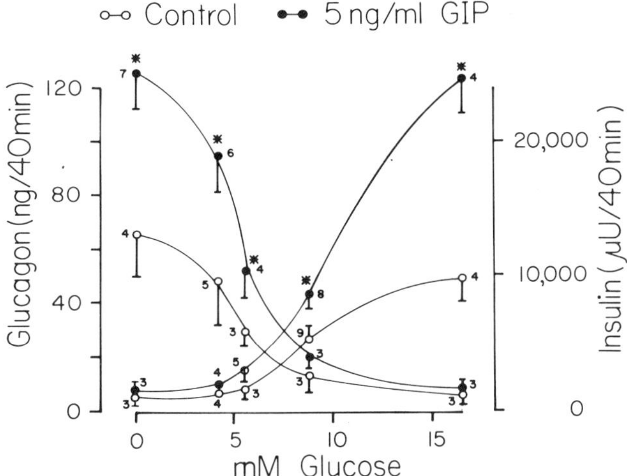

Fig. 18. The effects of graded glucose concentrations on the IR-glucagon and IRI release from the perfused rat pancreas in the presence (●) and absence (○) of 5.0 ng/ml GIP. IR-glucagon release by GIP was suppressed by increasing the glucose concentration

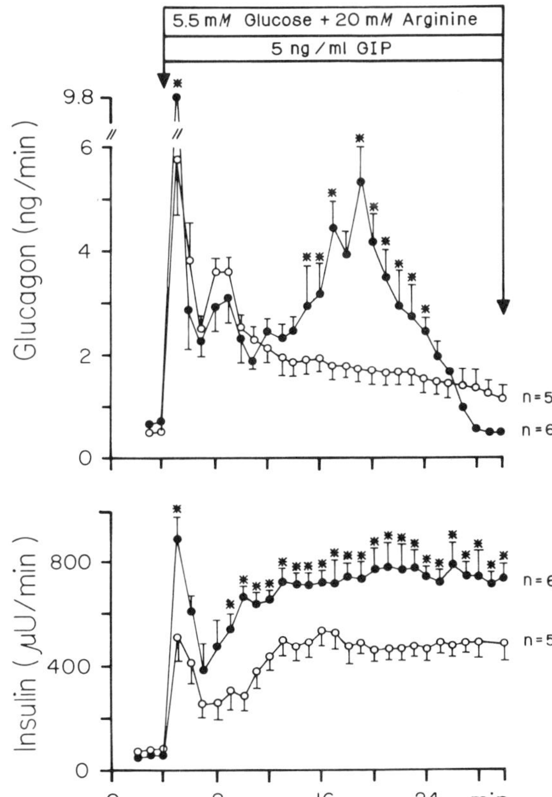

Fig. 19. The biphasic nature of the IR-glucagon response in the isolated perfused rat pancreas to infusion of 5.0 ng/ml GIP in the presence of 20 m*M* arginine can be seen. The second phase is transitory in nature

non-lipolytic and inhibitory for glucagon-stimulated lipolysis, but was without effect on lipolysis induced by secretin or vasoactive intestinal peptide (VIP). GIP could displace ^{125}I-glucagon bound to fat cells, identical to glucagon. VIP and secretin had been shown previously by Bataille et al. (1974) not to displace ^{125}I-glucagon. Ebert and Brown (1976) reported similar findings in that GIP demonstrated a weak lipolytic effect but a strong antilipolytic action on glucagon-stimulated lipolysis by selectively blocking the activation of adenylate cyclase by glucagon and to a lesser extent secretin. It would appear that GIP competed with glucagon for specific receptors on adipocytes but did not activate adenylate cyclase, therefore blocking the lipolytic action of glucagon.

Stimulation of lipoprotein lipase activity by GIP in 3T3-Ll cells, an established mouse embryo fibroblast line resembling an adipocyte has been described by Eckel et al. (1979). Cultured preadipocytes were incubated for 2 h at 37 °C with GIP concentrations ranging from 0.005 to 5.0 ng/ml. GIP increased lipoprotein lipase activity measured in both the culture medium and in acetone-ether extracts of cells. A dose-response relationship was strongest for the effect of GIP on the enzyme activity in extracts of acetone-ether powders of the cells. Eckel et al. (1979) considered that the increased lipoprotein lipase activity produced by GIP could provide a mechanism for clearance of the triglyceride of chylomicron after feeding.

c) Proinsulin Biosynthesis

Proinsulin biosynthesis as measured by ^{3}H-leucine incorporation and insulin secretion have been studied by Schäfer and Schatz (1979) in collagenase-isolated rat pancreatic islets. The islets had been incubated with 1.0 and 2.0 mg/ml glucose for 3 h in the presence of 5.0 μg/ml GIP. Both incorporation of ^{3}H-leucine and release of insulin were augmented by GIP. This insulinotropic activity was not observed when islets were incubated with 20 μg/ml of a mixture of 17 naturally occurring amino acids in the presence of 5 μg/ml GIP. Insulin release also was not augmented by GIP in the presence of the amino acid mixture when the glucose concentration was 1 mg/ml. These results with amino acids differed from those described by Fujimoto et al. (1978), who investigated insulin secretion from monolayer cultures of the islet cells of neonatal Wistar rats. They described an enhanced insulin response to GIP in the presence of amino acids. They also confirmed in this preparation the glucose-dependent nature of the insulinotropic action of GIP. Doses as low as 1.0 ng/ml significantly enhanced IRI release when 16.5 m*M* glucose was present, but at a glucose concentration of 1.7 m*M* a GIP concentration of 10 ng/ml or greater was required. A significantly increased IR-glucagon secretion which could not be suppressed by a high glucose concentration was also observed in this preparation.

Schäfer and Schatz (1979) indicated that insulin release was enhanced to a relatively greater extent than insulin-proinsulin biosynthesis. As indicated earlier, the collagenase-isolated islets did not respond with release of IRI until extremely high concentrations of GIP (10 μg/ml) were added (Schauder et al. 1975), whereas the monolayer cultures (Fujimoto et al. 1978) responded in a way similar to the perfused rat pancreas preparations of Pederson and

Brown (1976) to GIP concentrations as low as 1.0 ng/ml. These apparently contradictory pieces of work support the hypothesis that collagenase treatment alters receptor sites for GIP on the islet cell membrane, which will recover following short-term culture. The collagenase-isolated islet must be considered an unsuitable model for studies on the action of GIP and probably also on other insulinotropic peptides.

D. Radioimmunoassay

I. Development

1. Production of Antisera

Porcine GIP, in common with many small molecules, has been repeatedly shown to be a poor immunogen. In the production of antisera the peptide has been normally conjugated to bovine serum albumin (BSA) by the use of the carbodiimide condensation reaction described by Goodfriend et al. (1964). The radioimmunoassay described by Kuzio et al. (1974) utilized an antiserum raised in guinea-pigs to GIP conjugated to BSA by this technique. The preparations were injected subcutaneously into at least two sites on the lower abdomen at a dose of 25 nmol per immunization. First immunizations were performed using an emulsion with Freund's complete adjuvant. Repeated immunizations were carried out at 30-day intervals, with serum being taken 8-12 days after immunization. Sarson et al. (1980) immunized rabbits with 25 nmol porcine GIP conjugated by carbodiimide condensation to 12.5 nmol BSA, emulsified in Freund's complete adjuvant and injected into four subcutaneous sites. Booster injections of 20 nmol GIP conjugated with BSA and emulsified in Freund's incomplete adjuvant were given after 3 months, and three further booster injections were given at 4-month intervals using keyhole limpet haemocyanin instead of BSA as the protein carrier. The antisera raised by Morgan et al. (1978) was to porcine GIP conjugated to ovalbumin by the glutaraldehyde condensation technique of Reichlin et al. (1968). Rabbits were injected intradermally in multiple sites on their backs according to the method of Vaitukaitis et al. (1971). Animals were boosted intramuscularly after 5 months with 20 nmol conjugated GIP and intravenously at 7 months with approximately 1 nmol unconjugated peptide. Initial immunizations were with Freund's complete adjuvant (2 : 1). Lauritsen and Moody (1978) successfully raised antibodies to semipurified GIP prepared from a CCK-PZ-containing preparation which was 20% pure with respect to cholecystokinin activity. The immunogen was emulsified with an equal volume of Freund's complete adjuvant and multiple intradermal sites of injection were used.

2. Characterization of Antisera

The antiserum prepared by Kuzio et al. (1974) in guinea-pigs and subsequently employed by other groups was initially described as showing no significant cross-reactivity with motilin, natural porcine secretin, synthetic glucagon, synthetic human gastrin (15-Leu), pure CCK-PZ or GIP 1-14. VIP was later added to this list and Brown et al. (1975b) described the heterogeneity of

IR–GIP in human serum samples using this antiserum. In particular an IR–GIP of approximately 7500 daltons was observed. Sarson et al. (1980) described their antibody as showing negligible cross-reactivity with gastrin, CCK-PZ, VIP, glucagon, secretin, pancreatic polypeptide, insulin, neurotensin, somatostatin, substance P, urogastrone and motilin. A crude colonic extract was used to assess cross-reactivity with enteroglucagon and other neurohormonal peptides. Regional specificity of the antiserum to the GIP molecule was assessed using synthetic GIP fragments 1-16, 1-28, 15-43 and 26-43. This study revealed that a large portion of the C-terminal molecule was required for binding. A slight cross-reactivity (0.8%) of the antiserum with pancreatic glucagon was observed. This antiserum also recognized a large molecular weight IR–GIP. The antiserum of Morgan et al. (1978) showed negligible cross-reactivity with pancreatic polypeptide, synthetic human C-peptide and bovine insulin. The regional specificity of the antiserum was not determined, nor was there discussion of whether or not this antibody identified the 7500-dalton IR–GIP.

3. Preparation of ^{125}I-GIP

Kuzio et al. (1974) and Brown and Dryburgh (1979) described incorporation of ^{125}I into GIP when using the chloramine-T method of Greenwood et al. (1963). The iodination was performed in 0.4 *M* phosphate buffer at pH 7.5. Five micrograms of GIP was dissolved in 100 µl of 0.4 *M* PO_4 buffer, pH 7.5, in a siliconized 12 × 75 mm glass tube. To this was added 1.0 mCi of carrier-free ^{125}I (10 µl). Mixing was achieved by gentle bubbling. Forty micrograms of chloramine-T, in 10 µl buffer, were added and mixed thoroughly. The reaction was stopped after 15 s by the addition of 252 µg sodium metabisulphite (20 µl).

The reaction products were transferred to a 0.9 × 27.5 cm column of Sephadex G 25 fine which had been equilibrated overnight with 0.2 *M* acetic acid containing 2% Trasylol and 0.5% bovine serum albumin (Fraction V, Sigma). Fractions of 300 µl were collected, counted and tested for adsorption to charcoal and binding to antiserum in normal assay conditions. Fractions demonstrating the highest antibody binding and highest charcoal adsorption, usually on the descending limb of the labelled peptide peak, were pooled and stored at −20 °C in acid ethanol (100 ml 99% ethanol and 1.5 ml conc. HCl) at a final ethanol concentration of 50%. The average specific activity of the iodinated peptide was 90 µCi/µg. It was found to be stable under these conditions for approximately 30 days. ^{125}I–GIP with similar specific activities have been produced by Morgan et al. (1978) and by Lauritsen and Moody (1978). The former suggested performing the iodination in an ice/water bath and cooled all the colutions to less than 10 °C before use. They also reduced the concentrations of chloramine-T and sodium metabisulphite and shortened the exposure time to chloramine-T.

Sarson et al. (1980) used a lactoperoxidase method for ^{125}I incorporation into GIP, with partial purification on Sephadex G 50 superfine with 100 m*M* formic acid with 10 µ*M* HSA, 20 m*M* KI and 1000 KIU aprotinin per millilitre. Burhol et al. (1980) separated the iodination mixture on a Sephadex G 15 column (15 × 100 mm) followed by an SP-Sephadex G 25 (16 × 80 mm) eluted with 100 mM sodium acetate buffer, pH 5.0, containing 150 mM NaCl, 70 μM HSA and 3.08 mM NaN_3. The second stage of purification apparently produced a 2.5-fold increase in specific activity. Ebert et al. (1977) claimed an improvement in the sensitivity of the assay following minor modifications of the iodinating procedure. A much smaller quantity of GIP (500 ng) was used. The label was also repurified on the day of assay by treatment with 0.5 g Dowex (Cl^-, 50–100 mesh). The Dowex was added to the label aliquot, mixed and centrifuged for 5 min, thus removing ^{125}I which may have been liberated from the peptide. This repurification reduced the non-specific binding to 4%-6%; the displacement achieved by 6 pg was 15%, and 50% displacement could be obtained by 82 pg GIP.

Ross et al. (1977) used a ^{125}I-GIP purified by initial adsorption to microfine silicate (QUSO), stored in acid-ethanol at −20 °C and further purified prior to use on a Sephadex QAE 25 anion-exchange column. Elution from the column. was with 0.04 M Tris buffer, pH 8.6. Sensitivity down to 50 pg/ml was achieved.

4. Assay Procedure

The diluent buffer used throughout most of the published GIP radioimmunoassays was a 40 mM phosphate buffer, pH 6.5, stored as a stock solution of 400 mM at 4 °C and diluted 1 : 10 prior to use. Outdated blood-bank plasma which was charcoal extracted was added to a concentration of 5%, and Trasylol 7500 KIU per 100 ml was added. Morgan et al. (1978) substituted a human serum albumin preparation in their diluent buffer.

Assays were set up in a cold tray at 4 °C and all solutions were refrigerated prior to use. Incubations in the assay described by Kuzio et al. (1974) were of the equilibrium type and little difference could be observed between 48 h and 72 h incubation. As with any radioimmunoassay, conditions will vary with antibody.

In the assay technique of Kuzio et al. (1974), between 5 and 6×10^3 CPM of ^{125}I–GIP were used in each tube. An artificial control for inter- and intra-assay variability was established and separation was achieved with dextran-coated charcoal. A double antibody technique using donkey anti-rabbit γ–globulin has also been used. Lauritsen and Moody (1978) were able to achieve separation of ^{125}I–GIP and antibody bound ^{125}I–GIP with ethanol. O'Dorisio et al. (1976) separated bound and free components by precipitation of the bound component with 1.0 ml of 16% polyethylene glycol as described by Desbuquois and Aurbach (1971).

Most groups have used charcoal-extracted plasma in the diluent buffers. Burhol et al. (1980) examined plasma samples for IR–GIP, before and after charcoal treatment. They were unable to detect IR–GIP after charcoal treatment in plasma samples which had been shown, prior to treatment, to contain 1.2 ng/ml. Porcine GIP added to plasma samples to a concentration of 500 pg/ml was also completely removed by charcoaling.

II. Immunoreactive GIP (IR–GIP)

1. Release of IR–GIP

a) Response to a Meal

The radioimmunoassay for GIP has been used to study the physiological regulatory mechanisms for GIP release. Kuzio et al. (1974) reported that serum IR–GIP levels in normal human subjects following an overnight fast (12 h) were 237 ± 14 pg/ml (mean ± SEM) with a range of 75–500 pg/ml. After feeding a mixed meal, serum levels increased in a biphasic manner to approximately 1200 pg/ml within 45 min of food ingestion and remained significantly above basal for in excess of 4 h (Fig. 20). Burhol et al. (1980), using a different antibody and a highly purified ^{125}I–GIP, reported fasting plasma levels of 65 ± 10 pg/ml (mean ± SEM). Fasting plasma levels using a partially purified ^{125}I–GIP in the assay were found to be approximately 50% higher. In response to a test meal, IR–GIP levels increased to a peak of approximately 500 pg/ml within 60 min. The antiserum used in these studies recognized almost exclusively IR–GIP with a molecular weight of 5000 daltons.

Similar fasting levels and ranges for IR–GIP have been reported using the assays developed by Morgan et al. (1978) and Lauritsen and Moody (1978), and the former have reported an identical pattern of release following a mixed meal. Sarson et al. (1980) have reported absolute fasting values and meal-stimulated values which differ greatly from those of most other groups, being five times lower. There were, however, similarities in the degree of change observed in response to a mixed meal.

b) Response to Carbohydrate Ingestion

α) Stimulation

One of the necessary criteria which must be satisfied to establish a role for a peptide in the enteroinsular axis is the demonstration that it can be released into the circulation by ingestion of secretagogues known to release insulin, notably glucose. Cataland et al. (1974) followed serum levels of IR–GIP, IRI and glucose after ingestion of 75 g glucose. Mean fasting GIP levels were

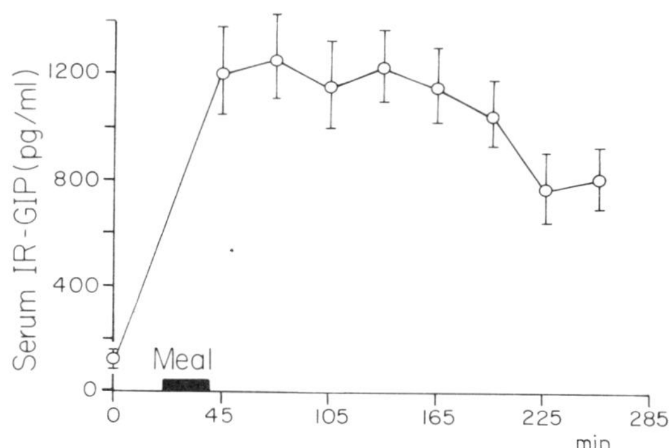

Fig. 20. Serum profile in response to a mixed meal, following an overnight fast, in six normal volunteers (mean ± SE)

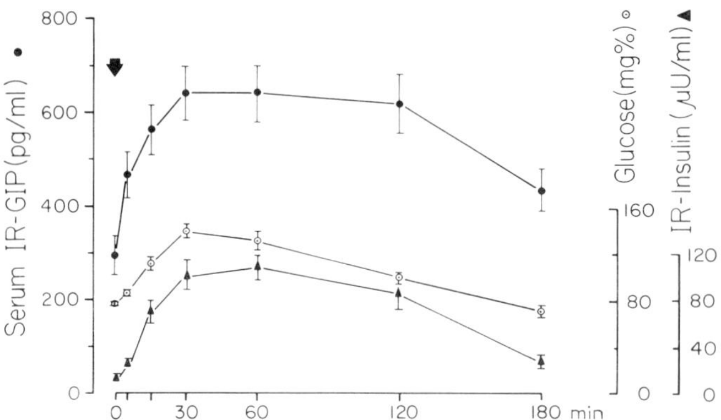

Fig. 21. Mean serum IR–GIP, IRI and glucose levels in 21 normal human volunteers following the ingestion of 75 g glucose

319 ± 18 pg/ml; these levels became significantly elevated within 15 min of glucose ingestion and peaked at 30 min at 747 ± 59 pg/ml. IR–GIP levels tended to remain at a peak between 30 and 60 min, declining slowly over 180 min (Fig. 21). Intravenous glucose (25 g) given as a single injection did not produce an increase in serum IR–GIP levels. Measuring portal venous levels of IR–GIP, IRI and glucose in response to oral glucose, these workers also demonstrated that portal GIP levels were significantly elevated within 2 min of glucose ingestion, whereas insulin levels were increased at 5 min. Cataland et al. (1974) also reported that several of their subjects demonstrated a fall in fasting levels of IR–GIP at 20–40 min after intravenous glucose, and they suggested that glucose, insulin or some other factor could bring about regulation of IR–GIP release by feedback inhibition.

Using the technique of hyperglycaemic clamp, Andersen et al. (1978) studied the release of IR–GIP after oral glucose and its association with insulin release in normal human volunteers. Blood glucose was maintained at 125 mg/dl above basal for a 2 h period using a primed continuous intravenous infusion coupled to a servo-controlled negative feedback system. After steady-state hyperglycaemia was established, glucose was ingested at a dose of 40 g/m^2 and the steady-state hyperglycaemia maintained. Little change in plasma IR–GIP levels occurred when intravenous glucose alone was administered, but following oral glucose IR–GIP levels rose from 305 ± 34 pg/ml (SEM) to a peak within 40 min of 752 ± 105 pg/ml. Plasma IRI levels followed a typical biphasic pattern to the induction of the square wave of hyperglycaemia. Oral glucose ingestion produced a highly significant increase in IRI levels (above those produced by intravenous glucose alone). When intravenous glucose alone was given, the mean IRI value between 90 and 120 min afterwards was 98.6 ± 10.1 μU/ml, as compared with 165 ± [illegible]5 μU/ml for oral + intravenous glucose; blood glucose levels were, however, similar (Fig. 22). These studies confirmed the work of McIntyre et al. (1964) in a situation in which blood glucose levels were rigidly controlled and in addition demonstrated that the time course of the rise of IR–GIP and IRI were nearly identical. Andersen et al. established a euglycaemic clamp in normal human volunteers to maintain basal blood glucose levels during a similar 2 h study, conducted to investigate the effect of oral glucose in a normoglycaemic situation. A primed continuous insulin intravenous infusion of 120 mU/m^2 per minute was given together with a servo-controlled glucose infusion. A hyperinsulinaemic condition (approximately 300 μU/ml) was produced. A similar dose of oral glucose was administered at time 60 min. Plasma IR–GIP levels were seen to be increased in a manner similar to those seen in hyperglycaemic clamp studies. However, if the plasma glucose level could be maintained at euglycaemia no increase in endogenous insulin release was observed, despite the rise in IR–GIP. In several instances blood glucose concentrations could not be adequately controlled and it was noted that if an increase of 20 mg/dl or more occurred, there was an increase in plasma IRI associated with the elevation in IR–GIP. They concluded that the effect of glucose-induced IR–GIP release on insulin secretion was dependent upon the prevailing degree of hyperglycaemia and was not inhibited in the presence of marked hyperinsulinaemia produced by exogenous insulin infusion. These studies did not rule out the possibility that other

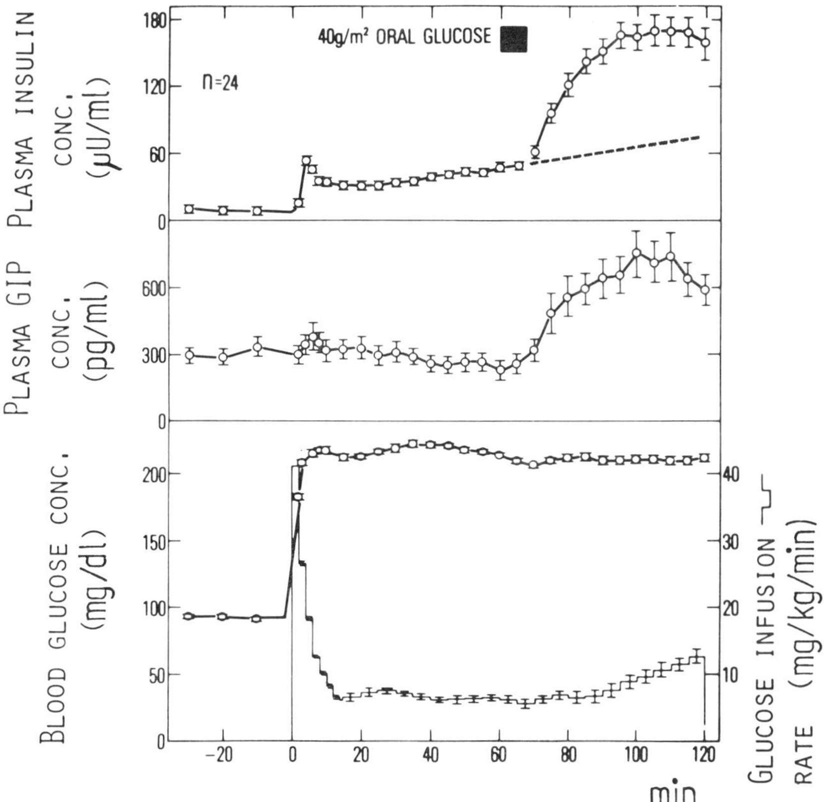

Fig. 22. Plasma IRI, plasma IR-GIP and blood glucose concentrations in an intravenous hyperglycaemic clamp situation with oral glucose in 24 normal volunteers. the *broken line* indicates the expected plasma IRI response for subjects receiving intravenous glucose alone

substances may be released by glucose ingestion and play a role in the mediation of the enteroinsular axis.

Pederson et al. (1975b) carried out studies in the dog to demonstrate that IR–GIP release occurred in a dose-related manner following oral glucose loads. Similar observations were made by Crockett et al. (1975) in man following oral administration of increasing amounts of glucose (Fig. 23).

Pederson et al. (1975b) also compared IRI release following (a) oral + intravenous glucose and (b) exogenous porcine GIP + intravenous glucose. Although the serum levels of IR–GIP achieved by the intravenous infusion of porcine GIP were higher than those achieved following oral glucose, the insulin output was also greater. Martin et al. (1975) concluded from studies in dogs with Mann-Bollman fistulae that both 10% and 20% solutions of glucose introduced directly into the duodenum were potent stimuli for the release of IR–GIP. A greater IR–GIP response was observed to 20% glucose, as could be expected. Equal volumes of 10% and 20% were used. IR–GIP release, however, was not observed when hyperosmolar solutions of galactose or mannitol were used. Martin et al. also concluded that the IR–GIP response to glucose was dose related and that duodenal osmoreceptors did not play a primary role in the physiological release of IR–GIP.

Cleator and Gourlay (1975) induced a significant increase in IR–GIP release after ingestion of 100 g galactose in 200 ml water, whereas Martin et al. (1975) reported that intraduodenal administration of 5 g galactose as a 10% solution in dog did not produce an increase in serum IR–GIP. This was not found to be so in man, when Morgan et al. (1979) studied the insulinotropic effect of 50 g galactose given orally to five normal volunteers, with and without concomitant hyperglycaemia. The hyperglycaemia was established by giving a bolus injection of 25 ml of 50% glucose and maintained by constant intravenous infusion of 10% glucose at a rate of 250 mg/kg per hour for 180 min. The 50 g dose of galactose was dissolved in 300 ml of chilled water (flavoured with lemon juice, pH 4.0) and administered 60 min after the experiment began. Ingestion of 50 g galactose without hyperglycaemia was followed by a rapid rise in plasma IR–GIP to a mean peak level of 900 ± 65 pg/ml at 30 min, with a small increase in IRI. A small rise in blood glucose is usually associated with galactose ingestion and Morgan et al. (1979) suggested that this slightly elevated blood glucose could be sufficient to sensitize the B cells to the IR–GIP

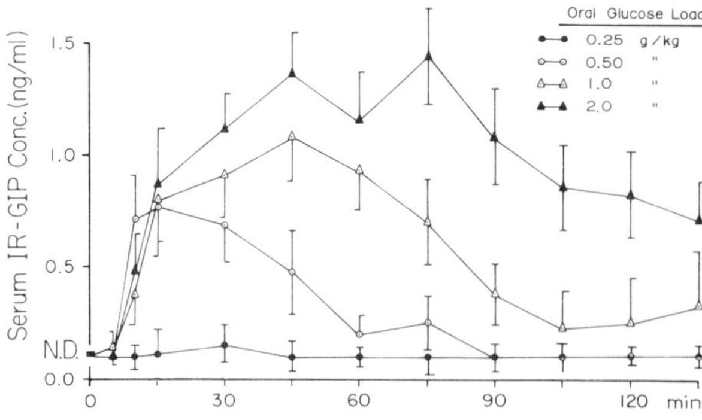

Fig. 23. Serum IR-GIP profiles in the dog following ingestion of increasing glucose loads

released by galactose and thus account for the modest increase in serum IRI observed when galactose alone was ingested (Marks and Samols 1969; Shima et al. 1972).

Morgan et al. (1979) showed that ingestion of 50 g galactose in the presence of hyperglycaemia was followed by a prompt large rise in IRI. They did observe a much lower plasma galactose level when it was ingested in the hyperglycaemic state. The mechanism behind this reduced galactose absorption could be one of several, including delayed gastric emptying, increased clearance due to the hyperinsulinaemia induced by the intravenous glucose (and released IR–GIP) and interference with galactose absorption by the higher insulin levels (Beyreiss et al. 1964). In the rat, Morgan (1979) has shown that IR–GIP release can be significantly stimulated by glucose, galactose and sucrose (Fig. 24). However, fructose was without effect. The same results were obtained in man and it was also reported that the plasma IR–GIP increase in response to sucrose was significantly delayed, relative to that seen with glucose.

β) Mechanism of Release of IR–GIP by Carbohydrate

The delayed secretion of IR–GIP in response to sucrose ingestion (Creutzfeldt et al. 1979; Morgan 1979) may be due to the fact that sucrose requires hydrolysing before exerting an effect. Morgan (1979) concluded that GIP secretion was dependent on the active transport of monosaccharides, and to support this statement cited the work by Crane (1965), who showed that a Na^+-dependent carrier protein was shared by both galactose and glucose for active transport. This carrier protein, however, has a lower K_m for glucose than for galactose. On the other hand, fructose was shown to be transported by a carrier mechanism independent of both Na^+ and the glucose transport system. Fructose transport is generally considered to be passive or facilitated (Crane 1968; Gray 1975).

A rat intestinal perfusion technique was used by Sykes et al. (1980) to investigate IR–GIP release in response to a variety of monosaccharides, monosaccharide analogues and disaccharides. They confirmed the ealier observations of Creutzfeldt and Ebert (1977) that 5 mMol phloridzin prevented intestinal absorption of glucose and abolished the release of IR–GIP.

Sucrose and maltose but not lactose stimulated IR–GIP release. The failure of lactose to do so was attributed to the low levels of lactase found in the

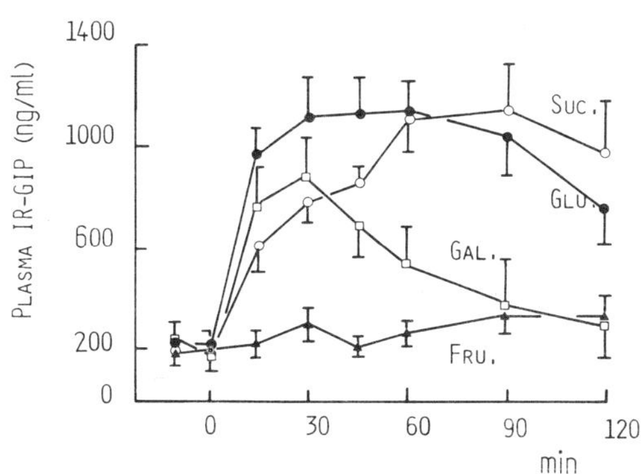

Fig. 24. Plasma IR-GIP responses to oral carbohydrate tolerance tests in normal human volunteers. Fifty grams of glucose, galactose of fructose or 100 g sucrose were ingested at time 0

intestinal mucosa of the mature rat. Glucose, galactose, 3-*0*-methyl glucose and α- or β-methylglucoside also produced significant increases in IR-GIP release. It was concluded from an examination of the molecular structures of the carbohydrates and analogues used that complete loss of ability to release IR–GIP resulted from the loss of the hydroxyl group at C 6, loss or rotation to the β form of the α-hydroxyl group at C-2 or loss of the pyranose ring confirmation. These observations also indicated that the basic molecular configuration for IR–GIP release agreed with the requirements for active transport (for hexoses) as postulated by Crane (1968). Hydrolysis of sucrose by a hydrolase situated in the intestinal brush border occurs before absorption can take place. The glucose released by the hydrolysis can be absorbed by a special glucose carrier which is Na^+ independent and closely associated physically with the enzyme complex (Ramaswamy et al. 1974). Excess glucose can, however, diffuse back into the intestinal lumen to be transported distally by conventional glucose carrier mechanisms to the site of digestion (Gray 1975). Ebert and Creutzfeldt (1980) demonstrated that when a 30 g glucose load was dissolved in 300 ml of 2.5% sodium chloride it produced a significantly enhanced IR–GIP and IRI response. This enhancement was not apparent when 100 g glucose was used as the stimulus.

The suggestion that absorption must take place to regulate IR–GIP release was first made following the observation that patients with coeliac disease and therefore marked malabsorption had a poor IR–GIP release following ingestion of a test meal (Creutzfeldt et al. 1976; Besterman et al. 1978). The number of immunofluorescent GIP cells was apparently normal (Creutzfeldt et al. 1976), suggesting that defective absorption might explain the diminished IR–GIP response. Treatment of rats with phloridzin (0.5 ml of 500 n*M* solution, 10 min earlier) most completely abolished the IR–GIP response to glucose administered by stomach tube, adding further support to the hypothesis that IR–GIP release is stimulated during the absorptive process and not simply by the presence of glucose in the small intestine (Creutzfeldt and Ebert 1977). IRI levels also showed no significant change.

TRIS (tris [hydroxymethyl] aminomethane) competitively blocks intestinal brush border sucrose. Ebert and Creutzfeldt (1978) showed that when normal subjects ingested a solution of 50 g glucose containing 5 g TRIS, IR–GIP and IRI release were significantly reduced from the control level. They claimed that the delayed monosaccharide absorption decreased IR–GIP release. Similar effects were described by Creutzfeldt et al. (1979) when they studied the effects of a glucosidase inhibitor (BAY g 5421) on serum IRI, IR–GIP and glucose levels following ingestion of 75 g sucrose. Goulder et al. (1978) have demonstrated in diabetics that carbohydrate absorption can be delayed by increasing dietary fibre. A decrease in insulin requirement occurs as a result. Creutzfeldt et al. (1979) investigated the effect of the addition of guar to a liquid test meal on the release of IR–GIP in normal human volunteers. The IR–GIP response remained unchanged although there was a significant lowering of serum glucose and IRI. These results were in contrast to the finding that IR–GIP release was dependent upon absorption and certainly warrants further investigation.

Sirinek et al. (1977) investigated the augmentation of IR–GIP release in response to intraduodenal glucose by intravenously administered gastrin, pentagastrin, CCK-PZ secretin and glucagon in dogs. In the control situation, 50 ml of 20% glucose solution was infused intraduodenally over a 10–min period. The peptide to be tested was given intravenously for a 30–min period prior to intraduodenal glucose and administration was then continued for 120 min after the glucose was given. Exogenous gastrin, pentagastrin, CCK-PZ, secretin and glucagon were without effect on fasting IR–GIP levels. There was a marked augmentation of IR–GIP response to intraduodenal glucose by gastrin, pentagastrin and CCK-PZ; gastrin having the greatest effect. Two readily apparent hypothetical explanations were offered. The ability of the three peptides to release HCl from the stomach was identical to their ability to augment glucose-stimulated IR–GIP release; however, these workers were unable to release IR–GIP by acidification of the duodenum. Therefore, they suggested that augmentation of glucose-stimulated IR–GIP release was not related to their gastric acid secretory activity but that the augmentation was probably related to the structural similarities of the peptides.

c) Response to Triglyceride

Brown (1974) described an increase in serum IR–GIP levels following the ingestion of 100 ml (66 g) of a corn oil suspension (Lipomul) in normal human volunteers. Triglyceride-stimulated serum IR–GIP levels reached a peak of approximately 700 g/ml at approximately 120 min after ingestion (Fig. 25). Corn oil (Lipomul suspension) was reported to contain 34%–62% linoleic acid, 19%–49% oleic acid, 8%–12% palmitic acid, 2.5%–4.5% stearic acid, 0.1%–1.7% myristic acid and 0.2%–1.6% hexadeconic acid, without any glucose or other carbohydrate (Falko et al. 1975). Falko et al. (1975) also studied the IR–GIP response to ingested emulsified corn oil in normal volunteers. They reported simular increases in IR–GIP to those described by Brown (1974) and also showed that there was no increase in serum IRI, glucose and non-esterified fatty acid concentrations. This study showed quite conclusively that endogenously released IR–GIP was not insulinotropic in the absence of hyperglycaemia.

O'Dorisio et al. (1976) infused intraduodenally, in dogs, 50 ml of a mixture of medium chain triglycerides via a Mann-Bollman fistula. The mixture had

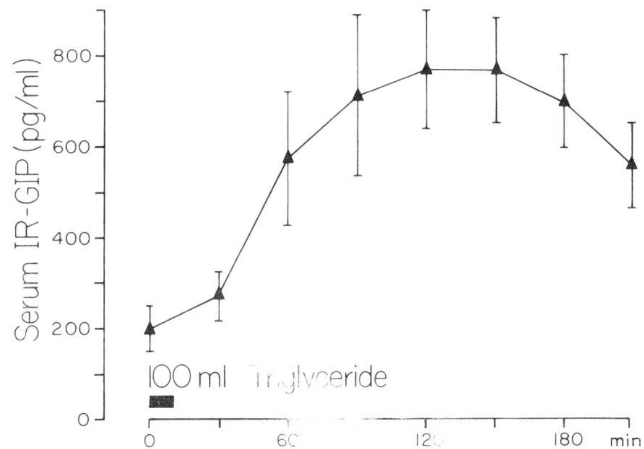

Fig. 25. Serum IR-GIP responses to the ingestion of 66 g triglyceride (100 ml corn oil) in normal human volunteers

a composition of 1.0% caproic acid (C_6), 75% caprylic acid (C_8), 23% capric acid (C_{10}) and 1% lauric acid (C_{12}). Medium chain triglycerides caused a modest increase in serum IR–GIP, which reached a peak concentration of approximately 500 pg/ml at 30 min and returned to pre–perfusion concentrations within 120 min. The response to intraduodenal instillation of medium chain triglycerides was much less than these workers had reported for corn oil (long chain fatty acids). Non-specific osmotic effects were excluded when it was shown that 3% saline or 20% mannitol administered intraduodenally did not evoke an increase in peripheral serum IR–GIP levels.

Ebert et al. (1976a), after studying IR–GIP release in response to a mixed high caloric test meal in patients with chronic pancreatitis, concluded that disturbed fat digestion may be responsible for the reduced release.

d) Response to Amino Acids

Brown (1974) reported that ingestion of a meat extract did not bring about a measurable change in serum IR–GIP. On the other hand, Thomas et al. (1976) were able to demonstrate that intraduodenal administration of a mixture of amino acids elevated serum IR–GIP concentrations in association with an increased IRI response. The response was of a transitory nature. Thomas et al. (1978) demonstrated quantitative differences with respect to the IR–GIP-releasing activity, insulinotropic activity and pancreatic exocrine stimulating effect of certain amino acids. They showed that a mixture of amino acids containing arginine, histidine, isoleucine, lysine and threonine caused a marked rise in integrated GIP and insulin secretion but only a small exocrine pancreatic effect as indicated by trypsin output. There was a transient increase in trypsin output, but this was considered to be a "washout" phenomenon and because it was not sustained they considered that CCK release was not produced by this mixture of amino acids. They were able to demonstrate, however, that a solution of amino acids containing methionine, phenylalanine, tryptophan and valine had only a minimal effect on IR–GIP and insulin secretion, whereas a substance with CCK-like activity was considered to have been released because there was a marked increase in bilirubin and pancreatic trypsin output.

Cleator and Gourlay (1975) were unable to produce a significant elevation in serum IR–GIP levels following ingestion of 280 g filet steak or 45 g of a meat extract. Thomas et al. (1978) considered that the differences between their studies and others could be attributed to differences in the concentration and quantity of amino acids present in the proximal small bowel. Low concentrations of the amino acids, 10 or 20 m*M*, in the small intestine demonstrated no IR–GIP-releasing effect.

III. Inhibition of IR–GIP Release

1. Insulin and Glucose

Brown et al. (1975b) studied the role of exogenous insulin in the regulation of IR–GIP release. Normal volunteers ingested a standard load of triglyceride which had been shown earlier to release IR–GIP without an associated increase in IRI. The change in serum IR–GIP was measured following ingestion of the triglyceride load, with and without an intravenous injection of insulin. When the insulin dose was administered, an intravenous infusion of glucose was also given to prevent hypoglycaemia. A bolus injection of insulin (7 mU) produced a transient elevation in plasma IRI and resulted in a significantly diminished plasma IR–GIP response to triglyceride ingestion. Brown et al. suggested that insulin was capable of inhibiting secretion of IR–GIP in normal man. Cleator and Gourlay (1975) described a similar suppression of triglyceride-induced IR–GIP release when an intravenous glucose infusion was given simultaneously, and Crockett et al. (1976a) described a 60% reduction in the integrated IR–GIP release in response to triglyceride during the intravenous infusion of 90 g glucose administered as a 10% solution at a constant rate of 0.5 g/min for 3 h. The peak response was also delayed.

None of these groups observed an IRI response to the ingestion of triglyceride, but intravenous glucose with simultaneous oral triglyceride produced a significantly greater IRI release than intravenous glucose alone. Serum glucose levels responded appropriately, being less well controlled when intravenous glucose was given alone. Crockett et al. (1976) concluded that it was possible that serum IR–GIP concentrations could be regulated by a negative feedback control mediated by either glucose or insulin. Sirinek et al. (1978) suggested that insulin was capable of attenuating the GIP response to oral glucose in a negative feedback fashion. In studies in which the plasma glucose level was clamped at 125 mg/dl above basal, Andersen et al. (1978) demonstrated that it was possible to release IR–GIP in response to ingestion of glucose. In this situation there was marked hyperglycaemia and moderate hyperinsulinaemia (endogenous IRI). In the presence of marked exogenously induced hyperinsulinaemia with controlled euglycaemia, ingestion of oral glucose still increased IR–GIP release significantly above basal but the effect was not significantly different from that observed with a hyperglycaemic clamp. Significantly reduced IR–GIP release following oral glucose was observed in the presence of marked hyperinsulinaemia with mild or moderate hyperglycaemia. Exogenously induced hyperinsulinaemia did not inhibit the IR–GIP response to oral glucose. Service et al. (1978) were unable to show any effect of hyperinsulinaemia induced by exogenous insulin on the basal secretion of IR–GIP in the absence of a change in plasma glucose concentration.

A feedback inhibitory control for IR–GIP release involving insulin is an expecially attractive mechanism. It would help to explain the elevated IR–GIP levels found in patients with diabetes and could also explain the higher levels found in patients with chronic pancreatitis (Botha et al. 1976; Ebert et al. 1976a). Ebert et al. (1976a) described significantly elevated IR–GIP release in response to a test meal in patients with chronic pancreatitis. The patients were divided into three goups on the basis of their insulin responses. A significantly higher IR–GIP release was observed in the group with the intermediate insulin and glucose responses than in those groups with the highest and lowest insulin and glucose responses. The group with the lowest responses had the greatest impairment of exocrine pancreatic function and therefore the greatest absorption problems.

2. Glucagon

Becker et al. (1973) and Hansky et al. (1973) described an inhibitory effect of glucagon on fasting- and food-stimulated gastrin levels in both normal individuals and patients with duodenal ulcer disease. Intravenous glucagon infusion, at a dose of 50 ng/kg per minute for 30 min, immediately followed by 500 ng/kg per minute for 30 min, has been performed in normal human volunteers. The effect of this dose of glucagon on fasting serum levels of glucose, IR-gastrin, IR–GIP and IRI was measured (Ebert et al. 1977). An insignificant depression of fasting IR–GIP levels was observed with the lower dose; this depression became significant after the dose was increased to 500 ng/kg per minute. The decrease in IR–GIP in the fasting situation was associated with an increase in IRI and serum glucose. Prior intravenous administration of exogenous glucagon at a dose of 50 ng/kg per minute completely suppressed the serum IR–GIP response to a liquid test meal in normal human volunteers. As in the studies with basal secretion of IR–GIP, the suppression of IR–GIP release was also associated with degrees of hyperinsulinaemia and hyperglycaemia in excess of those seen with the test meal alone. Discontinuation of the glucagon infusion resulted in an immediate increase in IR–GIP release above that seen with the test meal alone (Fig. 26) – a rebound phenomenon.

Ebert et al. (1977) discussed several possibilities as to the mechanism of the inhibitory action of glucagon on IR–GIP release. Delayed gastric emptying by glucagon and therefore removal of the secretagogues for IR–GIP release

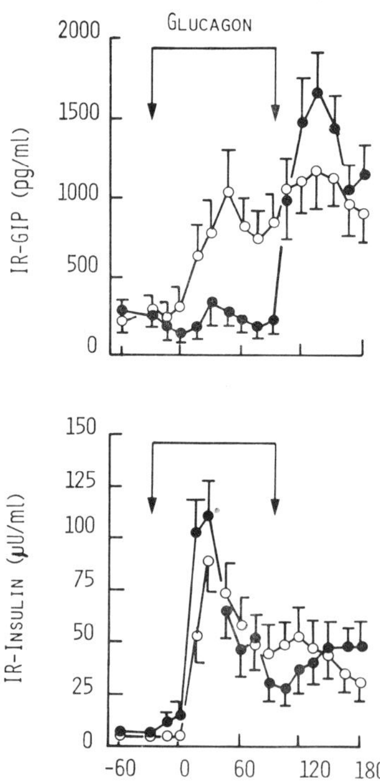

Fig. 26. Effect of glucagon (50 ng/kg per minute) on meal-stimulated IR-GIP and IRI release in six normal volunteers. Response to test meal alone (○) and with glucagon (●)

from the duodenojejunal region was excluded when, in a second series of experiments, the glucagon infusion was started 60 min after the ingestion of the test meal. Under these conditions a prompt and almost total inhibition of IR–GIP release occurred. Ebert et al. (1977) suggested that the rapid fall in fasting and test meal stimulated IR–GIP release, after the onset of glucagon infusion, was indicative of a direct effect on the GIP-producing cells of the gastrointestinal mucosa or modulation of release by changes in motor activity, assimilation of food and changes in serum levels of glucose or insulin. They argued that glucagon was not exerting its effect by influencing absorption, because the inhibition could be observed in the fasting situation. Hyperglycaemia, probably the main effect of glucagon, which was also observed in these studies, was considered not to be a cause of the inhibition because Creutzfeldt et al. (1976) and Ebert et al. (1976b) reported an exaggerated IR–GIP response to a test meal in patients with a pathological glucose tolerance, when glucose levels were far in excess of those reported in this study. Finally they suggested that only small rises in IRI in response to glucagon infusion were seen in the fasting state and no difference could be detected between the IRI release in response to the test meal with and without glucagon infusion. They concluded that the most probable mechanism by which glucagon inhibited secretion of IR–GIP was via a direct action at the cellular level.

3. Other Mechanisms

a) C-Peptide

The effect of exogenously administered rat C-peptide II on IR-GIP release in the rat intestinal perfusion model at a dose which achieved only three times the level found in the fed rat was described by Dryburgh et al. (1980). This dose of C-peptide totally abolished the IR–GIP response of the perfused rat intestine to fat stimulation. IRI levels remained unchanged throughout. C-peptide released after stimulation of the pancreas by 0.5 g/kg glucose and tolbutamide (250 mg/kg) administered intravenously in combination with insulin antiserum was also shown to significantly inhibit fat-stimulated IR–GIP release. They concluded that C-peptide was not a biologically inactive substance but was capable of exerting a controlling action on IR–GIP release.

b) Atropine

Larrimer et al. (1978) studied the effects of atropine on glucose-stimulated IR–GIP secretion in normal subjects. Glucose, 75 g over 60 min, was administered intraduodenally on two occasions. In the first instance intraduodenal glucose with concomitant infusion of saline was given as control, while in the test situation 15 μg/kg atropine was given intravenously as a bolus, followed by 17.0 μg/min for 60 min. Mean serum IR–GIP was significantly lowered by treatment with atropine and the integrated IR–GIP was significantly less. IRI levels were also reduced. Larrimer et al. concluded that atropine blunted the rise in IR–GIP seen with intraduodenal glucose via a direct effect on the gastro-intestinal tract which was other than inhibition of glucose absorption. In dogs, Baumert et al. (1978) demonstrated that subcutaneous injection of atropine sulphate (80 μg/kg) had no effect on basal IR–GIP release, but almost completely abolished the meal-stimulated rise in IR–GIP.

They concluded that the mechanism by which atropine suppressed IR-GIP release was unknown, but suggested that a delayed gastric emptying of the meal and a dependence of IR–GIP release on the vagus nerve were possibilities.

c) Somatostatin

In dogs, intravenous administration of somatostatin as a single, rapid injection at a dose of 3 μg/kg delayed the increase in serum concentrations of IRI, IR–GIP and glucose following the ingestion of 1.0 g/kg glucose (Pederson et al. 1975a). A 60–min infusion of somatostatin at a dose of 6.0 μg/kg per hour suppressed the IR–GIP response to oral glucose until cessation of the infusion. Serum glucose levels did, however, rise during the somatostatin infusion and by 60 min had achieved the same level as controls. They suggested that inhibition of glucose-stimulated IR–GIP release was not a result of the inhibition of glucose absorption. When somatostatin was injected before triglyceride ingestion, the mean IR–GIP serum levels were suppressed during the entire course of the experiment. The serum IRI levels were transiently inhibited by somatostatin injection, but then showed a rebound phenomenon. Somatostatin had been shown earlier by Alberti et al. (1973) to act at least in part directly on the pancreatic beta cell to produce inhibition. Pederson et al. (1975a) showed that pretreatment with somatostatin produced an 80% reduction in the peak IRI response to a dose of porcine GIP of a size to produce an insulinotropic effect in a normoglycaemic situation. This suggested a direct inhibition of the insulinotropic action of GIP at the pancreatic beta cell.

Creutzfeldt and Ebert (1977) studied the release of IR–GIP in man in response to a liquid test meal during the infusion of somatostatin. Somatostatin was administered at a dose of 250 μg/h, following a bolus injection of 125 μg. The IR–GIP reponse to the test meal was completely suppressed by somatostatin. IRI was also suppressed for as long as the somatostatin was infused, whereas blood glucose concentrations showed no difference, indicating, as Pederson et al. (1975a) had reported, that inhibition of IR–GIP release by somatostatin was not secondary to changes in glucose absorption.

IV. Nature of IR–GIP

1. Circulating Forms

Brown et al. (1975b) obtained serum samples from normal volunteers 45 min after the ingestion of 100 g glucose and 150 min after the ingestion of triglyceride. Serial dilution of these sera demonstrated parallelism with a standard curve in the radioimmunoassay. Serum samples subjected to gel filtration on Sephadex G 50 in 0.01 *M* phosphate buffer showed the presence of at least two immunoreactive components following ingestion of a mixed meal or corn oil. The greater part of the immunoreactivity eluted from the column at volumes corresponding to those obtained with ^{125}I–GIP, i.e. with a molecular weight of 5000 daltons. There was also an immunoreactive component with an apparent molecular weight of approximately 8000 daltons. In addition, an immunoreactive component was also observed to elute in the void volume.

Dryburgh (1977) extended these studies and showed that there was a difference between the IR–GIP response to corn oil and to glucose. After glucose ingestion (45 min) 50% of the total IR–GIP eluted as the 5000-dalton form, whereas after corn oil stimulation (90 min) only 28% eluted as this form. After glucose, 35% eluted as the 8000-dalton form as compared with 41% after corn oil. When porcine GIP (1 μg/kg per hour) was infused intravenously into dogs and serum samples were subjected to gel filtration on Sephadex G 50, a large percentage of the IR–GIP was found to elute in the void volume as well as in the expected 5000-dalton form. At 15 min all the IR–GIP was found in the void volume (Fig. 27). This void volume component could be significantly reduced by treatment of the samples with 6.0 *M* urea. Dryburgh (1977) suggested that the void volume IR–GIP represented a complex formed by the non-specific binding of GIP to a serum protein. Sarson et al. (1980) reported serum heterogeneity similar to that described in the earlier studies, although the total IR–GIP measured was much lower. Brown et al. (1979) assayed serum samples from a normal human volunteer, following the ingestion of glucose and triglyceride, with two different antisera raised in guinea-pigs. The antisera coded GP01 measured significantly less IR–GIP (Fig. 28) than GP24. Furthermore, Brown et al. (1980) could demonstrate no distinct pattern to the heterogeneity of the immunoreactive forms of GIP following ingestion

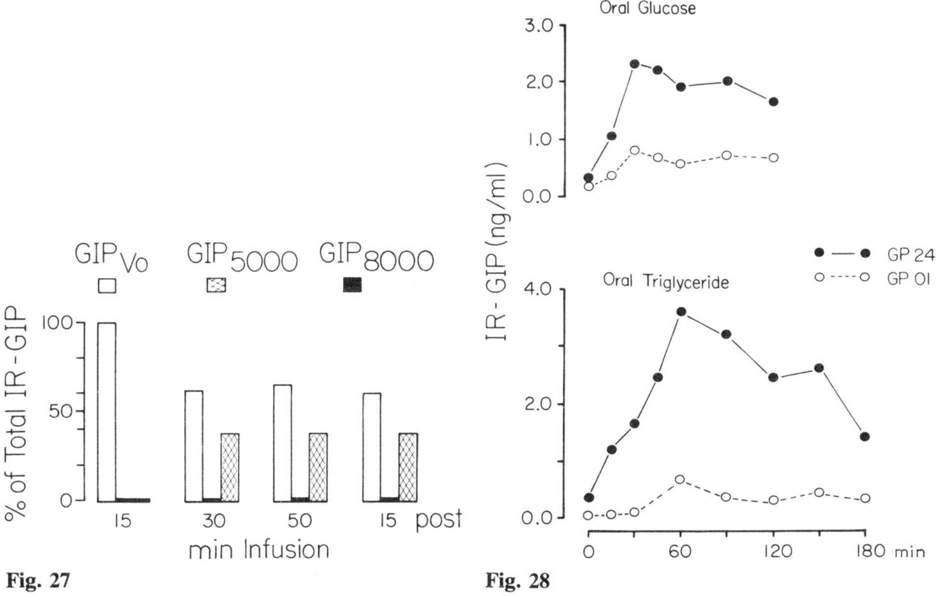

Fig. 27 **Fig. 28**

Fig. 27. Appearance of IR-GIP in the void volume of a Sephadex G-50 column following the intravenous infusion of porcine GIP in the dog, prior to the appearance of GIP_{5000}

Fig. 28. Release of IR-GIP in response to ingestion of glucose or triglyceride, as measured using two different antisera, GP01 and GP24, both of which were raised in guinea-pigs

of food substances. Interpretation of data from radioimmunoassay studies and comparison between laboratories using different antisera should be undertaken with extreme caution.

2. Tissue Forms

The heterogeneity of IR–GIP in tissue extracts has been reported by Dryburgh (1977). Chromatography of partially purified extracts of the duodenojejunal mucosa of pigs showed that they contained IR–GIP forms similar to those found in serum, i.e. with molecular weights of 5000 and approximately 8000 daltons. The highest ratio of IR–GIP 8000 : 5000 was found in a neutral soluble, methanol insoluble extract. The 8000-dalton form was described as labile. High-voltage electrophoresis and chromatography on carboxymethylcellulose have indicated that the larger form was less basic than the 5000-dalton material. Separation of the two immunoreactive species from the methanol insoluble solution on Sephadex G 50 has been described by Brown et al. (1978). A purification step involving ion-exchange chromatography on DEAE Sephadex G 25 was described by Brown et al. (1979), but to date the immunoreactive material has not been obtained in a pure form.

E. Localization

I. Cellular Localization

The gastro-intestinal extract from which GIP was eventually purified was made from the duodenojejunal region of the pig. Cellular localization using indirect immunofluorescence studies indicated that the polypeptide was present in cells situated predominantly in the mid-zone of the glands in the duodenum and upper jejunum in both dog and man (Polak et al. 1973). Cells demonstrating positive immunofluorescence using guinea-pig antiporcine GIP serum and fluorescein-labelled sheep anti-guinea-pig IgG were more numerous than secretin cells found in similar areas. Cytochemical and other staining reactions characterized the GIP cell as lead haemotoxylin-positive, Grimelius-positive, non-argentaffin and weakly positive for tryptophan. It was considered to belong to that series of cells referred to by Pearse (1969) as the APUD cells or endocrine polypeptide cells. The GIP cell was tentatively considered, as a result of electron microscopy, to be the D_1 cell because the distribution most closely matched the small granule-containing cell first described by Vassallo et al. (1971).

Mucosal samples of the small intestine from man, pig and dog were subjected to sequential or correlative silver impregnation techniques, immunocytochemical investigations and ultrastructural investigations by Buffa et al. (1975). They attempted to clarify whether or not a new cell recently isolated by Solcia et al. (1974) was a more likely candidate than the D_1 cell for the production of GIP. Immunofluorescence preparations demonstrated the presence of numerous medium-sized endocrine cells which reacted to anti-GIP sera in the duodenum and jejunum of all species. Restaining of the immunofluorescent preparations with the Sevier-Munger technique showed that the GIP cells stained strongly, whereas the Masson-Fontana technique and the diazo reaction were negative. The only other positive Sevier-Munger staining cell was the enterochromaffin cell (EC). However, this cell also stained positively with both the diazo and Masson-Fontana methods. In the Sevier-Munger stained preparations, a few weak yellow-brown staining cells which were neither GIP nor EC cells were also apparent. Small samples of tissue from all three species were fixed in a mixture of 2% paraformaldehyde and 2.5% glutaraldehyde in 0.1 *M* phosphate buffer at pH 7.3. Some specimens were post-fixed in 1% osmium tetroxide dehydrated in ethanol and embedded in Epon 812. Sections were stained with uranyl acetate and lead citrate. Cells corresponding to the K cell (Solcia et al. 1974) were revealed in all three species. They were recognized because of the double structure of many of their granules, comprising an electron dense (osmiophilic) core surrounded by a less dense matrix. The granules were round or slightly oval with an average diameter of 300 nm. One type of cell, when stained with the Sevier-Munger technique, corresponded exactly to the K cell seen by conventional electron microscopy. When examined by electron microscopy the Masson-Fontana preparations showed that the K cells were unreactive and that only the pleomorphic granules of the EC cells incorporated silver. Buffa et al. (1975) indicated that the immunofluorescent GIP cells were ultrastructurally indistinguishable from the K cells. Both failed to react with Masson-Fontana stain or diazo reactions and both stained intensely with the Sevier-Munger technique. They concluded that the K cell and the GIP cell were the same and that the D_1 cell was unrelated to GIP production.

Polak et al. (1975) used a quantitative histological and immunocytochemical approach to study the endocrine cells of the gastro-intestinal tract in genetically obese (ob/ob) mice and their heterozygous litter mates. They were able to demonstrate a hyperplasia of the GIP cells which was highly significant, but were unable to conclude whether it was primary or secondary. The ob/ob mice which were studied were between the ages of 9 and 12 months, which was actually the peak of their obesity. Polak et al. suggested that if the endocrine effect was primary then hyperplasia of the endocrine cells would be detectable at all ages, while if it was secondary then changes would only be seen in the obese period.

Distribution of IR–GIP in the small bowel wall of the hamster has been reported by Gaginella et al. (1978). After eversion of the entire small intestine, to the ileocaecal junction, the epithelial cells were isolated by a vibration technique. The remaining tissue was scraped to remove the rest of the villous components, leaving the muscle layer. All three layers were separately extracted with acid-ethanol, for GIP and VIP. The distribution of IR–GIP was quite different from that seen for VIP, in that IR–GIP could not be detected in muscle or nervous tissue, but was confined to the mucosal scrapings, where a concentration of 380 ± 50 ng/g tissue was detectable. VIP was found predominantly in the muscle layers.

An immunocytochemical technique was used by Smith et al. (1977) to demonstrate the presence of a GIP-like material in rat pancreatic islets. Pancreatic tissue was obtained from both neonatal and adult Wistar rats and fixed with Bouin's solution. The tissue was stained with a rabbit antiserum to GIP using the peroxidase-antiperoxidase technique. Tissue sections incubated with antiserum to GIP showed the presence of cells at the islet periphery. When adjacent sections were stained with glucagon antisera, a positive response was obtained over cells having identical positions in the islet to those reacting with GIP antiserum. The staining of the islet cells with GIP antiserum could be abolished by the addition of GIP. No positive reaction was observed when normal serum was used, but a positive staining could be obtained when the GIP antiserum was absorbed with glucagon. These findings indicated that the pancreatic A cells of neonatal and adult rats contained a GIP-like immunoreactive substance in addition to glucagon. An alternative explanation was also proposed, i.e. that GIP was taken up by these cells from the circulation.

II. Release Studies

Thomas et al. (1977) compared different areas of the small intestine for their effectiveness as releasers of IR–GIP. Glucose perfusions of duodenum, jejunum and ileum were carried out in normal human volunteers, using a four lumen polyethylene tube equipped with a proximal occluding balloon to prevent contamination by secretion and reflux. The distance perfused was 25 cm. Glucose-stimulated IR–GIP release was maximal in the upper small intestine and decreased with increasing distance from the pylorus. After infusion of glucose, the maximal serum IR–GIP concentrations achieved in the four areas perfused were: 1383 ± 152 pg/ml for the duodenum; 904 ± 87 pg/ml for the proximal jejunum; 545 ± 91 pg/ml for the mid-duodenum and 305 ± 38 pg/ml for the ileum. Peak serum IRI concentrations and integrated IRI secretion were also greater with perfusion of the duodenum of jejunum. The primary site for the endogenous release of IR–GIP was considered to be the proximal small intestine, but small quantities were also released from the distal small intestine. Release studies compared favourably with extraction studies.

F. Pathophysiology

I. Diabetes Mellitus

Identification of an insulinotropic action for GIP in man and development of a radioimmunoassay suitable for serum determinations led to a series of investigations into a possible role for the hormone in diabetes mellitus. Perley and Kipnis (1967) and Cerasi et al. (1973) had indicated that the enteroinsular axis was active or even hyperactive in maturity-onset diabetes. This later suggested to several groups a possible involvement of GIP, which has been considered to be a major component of the axis, in diabetic situations.

1. Non-insulin-Dependent Diabetes

Brown et al. (1975b) described the serum IR–GIP response to ingestion of 50 g glucose in normal and diabetic subjects. The patients ranged in age from 18 to 55 years, were within 20% of ideal body weight and had non-insulin-dependent diabetes which at the time of the study was being controlled by diet alone. Those subjects normally requiring oral hypoglycaemic agents had discontinued their medication for at least 3 days before the tests. The subjects were age and weight matched with normal volunteers and were receiving no medication. Ingestion of the glucose load following an overnight fast showed an IR–GIP response which was almost double that of the volunteers and a much reduced IRI response (Fig. 29).

In a group of 35 patients with maturity onset diabetes mellitus, Crockett et al. (1976b) demonstrated that ingestion of 75 g glucose produced a similarly elevated IR–GIP response to that described by Brown et al. (1975 b). Serum IRI responses were also observed to be delayed and lower than normal. The group of patients studied by Crockett et al. (1976b) had a mean obesity index of 1.44. The integrated IR–GIP response in the diabetics was 140.8 ng/min per litre and in the controls (mean obesity index, 1.11), 64.6 ng/min per litre. Despite the elevated serum IR–GIP responses in the diabetic subjects, no significant difference in absolute serum insulin levels between the two groups was observed. Insulin secretion was, however delayed and inappropriately low for the corresponding serum glucose level.

Ross et al. (1977) compared IR–GIP release during a 50 g oral glucose tolerance test in patients with maturity onset diabetes mellitus and in normal volunteers. The diabetic subjects were non-obese and treated by diet alone at the time of the study. If the patients were taking oral medication, this was discontinued for a period of 7 days prior to the text. The diabetic subjects had a mean elevated fasting plasma glucose level of 155 mg/dl and exhibited marked intolerance to the glucose load. In response to the glucose load, the plasma

IRI levels in the diabetic subjects demonstrated a much slower rise and a relatively delayed and diminished maximum secretion. The plasma IR–GIP release in the diabetic group was more rapid in onset and reached higher levels, and the integrated area of change was 2.15 times greater than seen in normal subjects. It was suggested that the exaggerated rise in plasma IR–GIP following ingestion of glucose in the diabetic subjects could be related to the diminished insulin response, supporting a previous suggestion that insulin may inhibit secretion of IR–GIP. Ross et al. further proposed that since GIP had been shown to be glucagonotropic, the hormone could contribute to the hypersecretion of glucagon, which has been shown to occur in diabetes.

Creutzfeldt et al. (1976) reported the composition of a provocative liquid test meal for maximal IR–GIP release. The standard meal, subsequently used on numerous occasions by Creutzfeldt and colleagues, contained 1031 cal per 550 ml and had an osmolarity of 1280 mosmol/litre. Creutzfeldt and Ebert (1977) studied IR–GIP, IRI and serum glucose release in response to the test meal in a group of 29 maturity onset diabetics (mean obesity index 1.41) and compared the responses with those in 25 normal volunteers. The mean fasting serum glucose value of this group was 130 mg/dl rising to over 400 mg/dl following ingestion of the test meal. This group of diabetics demonstrated a significantly elevated serum IRI response to the test meal with a concomitant hypersecretion of IR–GIP. The mean peak of IR–GIP response was in excess of 4.0 ng/ml as compared with approximately 1.0 ng/ml in the normal volunteers. They concluded that obese maturity onset diabetics had an exaggerated IR–GIP response to a high caloric text meal. The fasting and stimulated hyperglycaemia and the stimulated hyperinsulinaemia obviously did not contribute to feedback inhibition in these diabetic patients. Creutzfeldt and Ebert therefore suggested that the cells releasing IR–GIP in maturity onset diabetes were unresponsive to the proposed feedback inhibition by insulin or glucose.

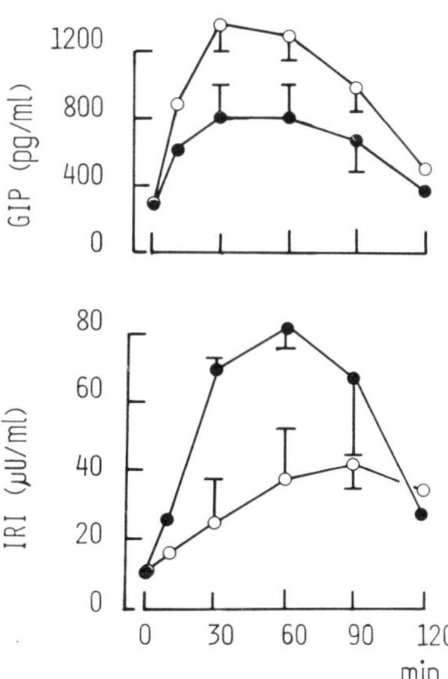

Fig. 29. Comparison of serum IRI and IR-GIP in nine normal subjects (●) and 16 diabetics (○) following ingestion of 50 g glucose

May and Williams (1978) investigated a group of diabetics with even milder glucose intolerance than those studied earlier, using either an oral glucose load of 75 g or a triglyceride load of 50 ml corn oil. The corn oil was used in these investigations because of a preliminary report by Brown et al. (1975b) that patients with maturity-onset diabetes mellitus with fasting glucose concentrations of 150 mg/dl or above have hypersecretion of IR–GIP in response to oral glucose but a normal response to emulsified corn oil. No enhancement of IR–GIP secretion was found in this group with mild diabetes in response to either a glucose or triglyceride challenge. When the obese subjects in this group were assessed independently from glucose intolerance, there was an association of obesity with higher integrated IR–GIP secretion in response to glucose but not to corn oil, confirming the earlier studies of Brown et al. (1975b).

It appeared from published data that there was some interrelationship between circulating levels of IR–GIP and the degree of glucose intolerance concomitantly present in mild forms of non-insulin-dependent diabetes. Creutzfeldt (1979), however, concluded, that the primary defect of the B cell, leading to delayed and inappropriate insulin release in maturity onset diabetes, could not be overcome by endogenously released IR–GIP. Moody (1977) suggested that there was insufficient evidence to propose that 'incretins' other than GIP existed, and Ebert et al. (1978, 1979) tested this suggestion with an immunological approach. They showed that passive immunization with a potent GIP antiserum completely abolished the initial incretin effect of intraduodenally administered HCl in anaesthetized rats which had simultaneous intravenous glucose infusion. The insulinotropic action of intraduodenal glucose was reduced but not abolished by GIP antiserum, but the results of oral glucose tolerance tests in conscious rats were not altered by prior injection of GIP antiserum. The insulinotropic action of exogenous GIP in the presence of antiserum was not tested. Creutzfeldt (1979) concluded that there were other incretins awaiting isolation and that a type of diabetes related to the absence of an incretin should not be ruled out.

Cataland and O'Dorisio (1980) reviewed the possible mechanisms which might account for the exaggerated IR–GIP secretion found in maturity onset diabetes. Manipulation of diet, by starvation for 3 weeks followed by refeeding for 3 days, nearly normalized the exaggrated IR–GIP response to a high caloric liquid test meal (Creutzfeldt 1979). A similar normalization was reported following a 3-week treatment with glibenclamide (Creutzfeldt and Ebert 1977). When tolazamide, without dietary restriction, was given to maturity onset diabetics, Cataland and O'Dorisio (1980) reported no change in serum IR–GIP levels although there was a significant reduction in basal and postmeal serum glucose levels in association with elevated IRI concentrations.

Ross and Dupré (1978) studied IR–GIP release following ingestion of triglyceride or galactose in a group of normal weight, non-insulin-dependent diabetics with a mean fasting plasma glucose value of approximately (130 mg/dl). Basal levels of IRI, IRG (glucagon) and IR–GIP in the plasma of the diabetics were found not to be significantly different from those in the plasma of normal subjects. After ingestion of 66 g triglyceride there was a rise in plasma IRG in the diabetic subjects which was not seen in the controls. No difference in IR–GIP reponse could be observed between the two groups. However, when intravenous glucose was given after oral triglyceride, the plasma IR–GIP response in normal subjects was suppressed, whereas in the diabetics there was no fall in plasma IR–GIP after intravenous glucose. Rises in plasma IR–GIP were observed in both normal and diabetic subjects following ingestion of 50 g galactose. Mean plasma IR–GIP was significantly higher in the diabetic subjects only at 10 min after ingestion. IRI levels were also comparable between the two groups. The effect of oral galactose on the response to intravenous glucose in the normal subjects was to produce enhancement of plasma IRI levels and improvement of glucose tolerance. In the diabetics, oral galactose, unlike oral triglyceride, also resulted in enhancement of the early rise of plasma IRI after intravenous glucose, but this was not enough to affect glucose tolerance. Ross and Dupré (1978) concluded that although their results suggested there was a partial resistance to the insulinotropic action of GIP in diabetes, the hormone could participate in the enteroinsular augmentation of insulin release demonstrated in maturity onset diabetes. These studies also suggested that the rise in plasma IRG seen in diabetics after oral triglyceride indicated the probable importance of intestinal factors in abnormalities of glucagon secretion after ingestion of nutrients not having a direct effect on the alpha cell. A role for GIP was considered to be compatible with the data presented.

Brown and Otte (1979a), after reviewing the literature on the pathophysiology of GIP, concluded that a physiological role for the hormone in the entero-insular axis had been established unequivocally, but a causal relationship between hypersecretion of insulin and elevated IR–GIP levels remained equivocal, although conceivable.

2. Insulin-Dependent Diabetes

Creutzfeldt and Ebert (1977) observed that fasting serum levels of IR–GIP were considerably elevated in untreated juvenile diabetics when compared with normals or a group of untreated maturity onset diabetics (mean obesity index, 1.41). The mean fasting serum levels in the juvenile diabetic group (with significant ketonuria and later development of insulin dependency) were 1693 ± 241 pg/ml as compared with control values of 266 ± 68 pg/ml and 521 ± 189 pg/ml in the maturity onset diabetics. Creutzfeldt and Ebert (1977) and Willms et al. (1978) observed that similar high fasting IR–GIP levels developed in a group of obese individuals following complete starvation. Serum IR–GIP levels in excess of 1500 pg/ml were observed within 72 h of food withdrawal. In contrast, serum IRI levels fell from a mean of approximately 20 μU/ml to approximately 6 μU/ml over the same period. Willms et al. (1978) concluded that the increased basal IR–GIP levels in both untreated juvenile diabetics and following starvation could have occurred because of the low insulin levels of the diabetics and the fall in insulin levels after starvation. The end result would be to remove a feedback inhibition control between GIP and insulin. An alternative suggestion in which ketone bodies would act directly on the GIP-producing cells was also proposed.

In a group of 14 untreated juvenile diabetics, ingestion of a liquid test meal was shown by Ebert and Creutzfeldt (1977) to produce no significant increase in IR–GIP secretion over and above the elevated fasting serum levels. Lack of significance was due to the large variation in individual responses. However, they did report that insulin treatment for 5 days normalized the high fasting serum IR–GIP levels, and that the IR–GIP response to the liquid test meal, in the absence of insulin, became normal. Reynolds et al. (1979) studied the serum IR–GIP response in insulin-dependent diabetics following ingestion of 1 g glucose per kilogram body weight. The fasting serum IR–GIP levels were lower than those observed by Ebert and Creutzfeldt (1977). However, these diabetic subjects were receiving prior treatment with insulin and in preparation for the test they received four daily injections of regular insulin, the last dose being given 9 h before the test. In response to the oral glucose load, the IR–GIP response was diminished as compared with that in normals. There was almost no IRI response in these subjects. These results compare favourably with the data of Ebert and Creutzfeldt (1977), who found that insulin treatment for 5 days normalized the elevated fasting serum levels of IR–GIP and the response to a liquid mixed meal.

Creutzfeldt et al. (1980) studied the effect of glucose and insulin on triglyceride and glucose-induced IR–GIP release in a group of 40 insulin-dependent diabetics. The evening before the test, insulin was withheld and the carbohydrate content of the supper and bedtime snack was reduced by 50%. The fasting serum levels of IR–GIP described in this study were higher than previously reported in well controlled diabetics and it was felt that this was as a result of withholding the intermediate-acting insulin dose the evening before the test. This supports the suggestion that basal values of IR–GIP are related to the serum insulin or blood glucose levels. Triglyceride (100 g)-induced IR–GIP secretion in the diabetics could be inhibited by exogenous insulin in the presence of moderate hyperglycaemia. No significant difference between the IR–GIP responses could be observed when intravenous glucose was administered with oral triglyceride and compared with triglyceride alone. It was concluded that because of the complete absence of endogenous insulin release in response to intravenous glucose in these subjects, as measured by radioimmunoassay for C-peptide, inhibition was due to insulin and not glucose. The inhibitory effect of IR–GIP release was present only briefly after the cessation of the insulin infusion. A rebound phenomenon was noticed when the infusion was stopped. When 100 g oral glucose was ingested to release IR–GIP it was found that the mean serum levels were similar, whether or not insulin was infused concomitantly. Creutzfeldt et al. (1980) considered that it was unlikely that this differential effect of insulin was due to differences in molecular forms of IR-GIP being released by the two secretagogues.

II. Pancreatitis

Botha et al. (1976) investigated the release of IR–GIP in response to glucose ingestion in patients with chronic pancreatitis. It had been shown previously that the IRI release following intravenous glucose was markedly impaired in chronic pancreatitis (Kalk et al. 1974) whilst the insulin response to oral glucose was relatively unimpaired (Bank et al. 1968; Raptis et al. 1971). CCK–PZ had been shown to stimulate insulin release when given alone and to enhance release induced by glucose in patients with chronic pancreatitis and in animals with experimentally induced pancreatitis. The CCK–PZ preparations used have since been shown to contain GIP.

The patients involved in this study were diagnosed as having chronic pancreatitis on the basis of criteria suggested by Bank et al. (1963), were all male and weighed between 43 and 79 kg. The patients were without cirrhosis or portal hypertension; they had a history of excessive alcohol intake, but no family history of diabetes. Both the subjects and the controls ingested 50 g glucose in 150 ml water. There was an excessive IR–GIP response to oral glucose in the patients similar to that shown in maturity onset diabetes. The plasma IRI levels were lower than in controls. It was suggested that the elevated levels of IR–GIP may be an important factor in the maintenance of relatively normal insulin responses to oral glucose in pancreatitis. These studies presented evidence of a further similarity between acquired diabetes and genetic diabetes in that IR–GIP responses were similar in both, fulfilling the postulate made by Vinik et al. (1974).

Ebert et al. (1976a) investigated IR–GIP release in a much larger group of patients with chronic pancreatitis, following administration of a test meal. The patients were subdivided into groups according to the degree of exocrine and endocrine insufficiency. As a group, the IR–GIP and glucose responses in the patients with chronic pancreatitis were significantly larger than in the controls. The IRI response, however, was significantly smaller, and IRG response was not significantly different from normal. An analysis was made of the relationships between the degree of exocrine and endocrine insufficiency and the IR–GIP response. No difference could be observed between the patients with mild steatorrhoea (<10 g/24 h) and those with severe steatorrhoea (>20 g/24 h). The IRI and serum glucose responses showed significant differences between the two groups. The patients with severe maldigestion had a much smaller IRI response and a much higher serum glucose level. When the patients were divided into subgroups based on their integrated IRI response to the test meal as an indication of beta cell secretory capacity, an inverse relationship between the glucose response and the IRI response was observed. Similar integrated IR–GIP responses were observed in the patients with a near normal IRI response and those with a response below 2.5 mU/ml per 180 min, the overt diabetics who were mostly insulin dependent. In those patients with a moderately impaired IRI response, a significantly higher integrated IR–GIP was reported (Fig. 30). Ebert et al. concluded that malassimilation of fat could not be the reason for the exaggerated IR–GIP response in patients with chronic pancreatitis, nor could the elevated serum glucose levels because intravenous glucose does not release IR–GIP. They considered that

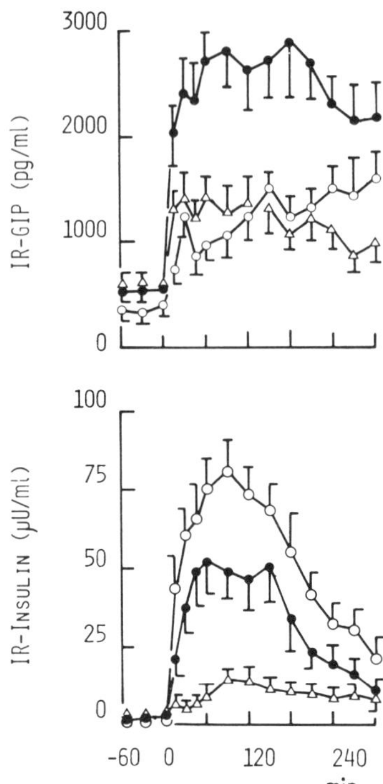

Fig. 30. Increase in serum levels of IR-GIP and IRI in 29 patients with chronic pancreatitis. The patients were divided into three groups according to their IRI response to a test meal. Group 1 (○) had integrated IRI levels of >7.0 mU/ml. 180 min (n = 11); group 2 (●) had integrated IRI levels of 2.5-7.0 mU/ml. 180 min (n = 11) and group 3 (△) had integrated IRI levels of <2.5 mU/ml. 180 min

loss of feedback control of IR–GIP release by insulin would explain the elevated levels both in patients with diabetes as well as in those with chronic pancreatitis. Both exocrine and endocrine functions are found to be impaired in patients with chronic pancreatitis, and therefore it was to be expected that those with the severest endocrine impairment might also have the highest degree of malassimilation of fat. In such a situation release of IR–GIP would be reduced so that increases due to loss of feedback inhibition as a consequence of endocrine insufficiency would not be apparent.

Ebert and Creutzfeldt (1980) tested the hypothesis that the major determinant for the release of IR–GIP in chronic pancreatitis was the degree of exocrine insufficiency. The release of IR–GIP in nine patients with massive steatorrhoea was tested following ingestion of 100 g fat with and without administration of enzyme substitute (9.0 g pancreatin). Pancreatin by itself did not induce a release of IR–GIP, but significantly greater serum concentrations were attained when pancreatin was added than when fat was given alone. Intravenous infusion of 0.7 g/kg glucose per hour was given to patients with pancreatitis, and measurements were made of the IR–GIP and IRI release in response to a triglyceride load, with and without oral enzyme supplement. The insulin response to the intravenous glucose load plus triglyceride ingestion was almost doubled when enzyme supplement was given, but as a result of the increased IRI release, the serum levels of IR–GIP were found to be reduced by 25%. This was considered to be another indication of the existence of the proposed feedback inhibition of IR–GIP release by insulin.

The existence of the postulated feedback mechanism between IR–GIP and IRI release has been investigated by Botha et al. (1978) in patients with insulin-requiring chronic pancreatitis, before and after withdrawal of insulin therapy. The provocative challenge for IR–GIP release was ingestion of 50 g glucose. These studies confirmed earlier reports that basal levels of IR–GIP were elevated in patients with chronic pancreatitis and became even more so following insulin withdrawal. After resumption of insulin treatment the basal IR–GIP concentrations returned to control levels. When the patients were on short-acting insulin, withdrawal from treatment the previous night resulted in higher basal IR–GIP levels than in those on insulin lente, probably due to continued effect of the previous night's dose. The IR–GIP response to ingestion of 50 g glucose was unchanged in both situations, although there was a tendency for IR–GIP levels to be higher in patients who were taken off the short-acting insulin preparation.

Ebert et al. (1976b) had earlier reported that untreated juvenile-onset diabetic patients also had elevated fasting IR–GIP levels and that the response to a test meal was also no different from normals. The severely insulinopoenic pancreatitic patients studied by Botha et al. (1978) had a smaller integrated IR–GIP response and lower peak plasma levels than the group of patients with pancreatitis studied earlier, who had better insulin responses. The apparent discrepancies could not be adequately explained, but it was suggested that they may be due to differences in small bowel pathology.

Creutzfeldt et al. (1976) studied IR–GIP release in patients following partial duodenopancreatectomy necessitated by complications due to chronic pancreatitis. Patients with chronic pancreatitis were taken as controls. There was a significantly elevated serum IR–GIP response in the patients after duodenopancreatectomy, which, it was suggested, could be partially explained by either the rapid passage of the test meal through the gastrojejunostomy into the small intestine or reduced feedback inhibition of IR–GIP secretion caused by the decreased IRI release. The latter possibility, however, was not supported by these studies.

Maintenance of elevated serum IR–GIP levels following duodenectomy demonstrated further the presence of GIP in the jejunum.

III. Other Gastrointestinal Disorders

1. Coeliac Disease

Ingestion of 550 ml of a liquid test meal containing 400 ml Dextro O.G.T. (18 g glucose, 14 g maltose, 12 g maltotriose, 56 g oligosaccharides), 100 ml 30% cream and 100 g Molico Instant (36 g protein and 52 g lactose), amounting to 1,031 cal and with an osmolarity of 1,280 mosmol/litre was used by Creutzfeldt et al. (1976) to study IR–GIP release in coeliac disease. In six patients with coeliac disease, IR–GIP hardly increased in response to the test meal. Biopsy material from the jejunal mucosa of four patients with coeliac disease was compared with normal tissue. The jejunal mucosa of the patients with coeliac disease revealed subtotal villous atrophy. The mucosal glands were well-preserved and contained numerous immunofluorescent GIP cells. This ruled out the possibility that the reduced IR–GIP response in coeliac disease was due to a reduced number of cells in the atrophic mucosa of the endocrine-producing areas of the small intestine. A suggested explanation was that in patients with coeliac disease, absorption is defective, and in the patients studied by Creutzfeldt et al. (1976), malabsorption was present at the time of investigation. Therefore if IR–GIP release was dependent upon the rate of absorption, reduced release could be expected.

Besterman et al. (1978) studied the effect of a mixed meal as a provocative challenge in patients with treated and untreated coeliac disease. They observed an impaired release of IR–GIP with the untreated disease, which did not normalize despite improvement of jejunal histology following adherence to a gluten-free diet. The reduced insulin levels in patients with untreated coeliac disease could be related to the low IR–GIP levels seen in these patients.

2. Duodenal Ulcer

Many possible mechanisms have been postulated to account for the gastric hypersecretion that is generally associated with duodenal ulcer disease.

Amongst these is the speculation that there may be an impairment of the release of an inhibitor of acid secretion from the small intestine. Cataland et al. (1977) compared IR–GIP release in 16 patients with radiographically and endoscopically proven duodenal ulcer disease. Serum IR–GIP levels were followed after the ingestion of a mixed meal and compared with serum values obtained from a group of normal volunteers. Similar fasting serum IR–GIP levels were obtained, but postprandially the mean levels in the duodenal ulcer patients were significantly greater. This observation suggested that a reduced release of IR–GIP was not contributing to the hypersecretion of gastric acid seen in duodenal ulcer patients.

Brown et al. (1975a) had indicated previously that neither endogenous acid nor acidification of the duodenum would stimulate the release of IR–GIP, so Cataland et al. (1977) concluded that it was unlikely that the elevated IR–GIP levels occurred as a result of gastric hypersecretion.

Buchanan et al. (1967) and Humphrey et al. (1972) have suggested that the enteroinsular axis appears to be overactive in chronic duodenal ulcer disease, in that greater than normal amounts of insulin were released following glucose ingestion. The IRI response to intravenous glucose, however, was found to be the same in both ulcer and control groups. A role for GIP in the hyperinsulinaemia seen with duodenal ulcer was suggested by Cataland et al. (1978). Ingestion of 50 g glucose produced significantly greater IRI and IR–GIP responses in patients with duodenal ulcer disease than in normal controls. The total integrated serum glucose concentrations were similar in both groups, thus excluding glucose intolerance. They were unable to identify the cause of the exaggerated IR–GIP release, but suggested that because patients with uncomplicated ulcer disease have a rapid gastric emptying rate, a faster rise and a higher peak of glucose concentration would result. The greater rate of glucose pre-

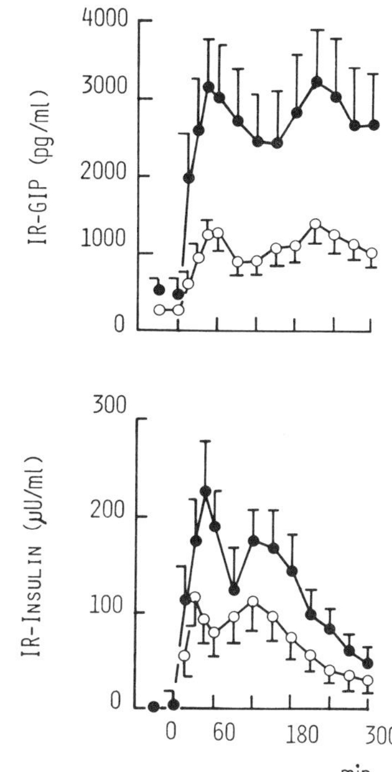

Fig. 31. Increase in serum IRI and IR-GIP responses in control subjects (○) and patients with duodenal ulcers (●)

sentation to the small intestine could be responsible for the augmented IR–GIP release.

Lauritsen and Moody (1978) compared insulin responses to intravenous and oral glucose challenges in both control subjects and patients with duodenal ulceration confirmed by radiological and endoscopic examination. None of the subjects had a diabetic oral glucose tolerance and mean fasting glucose concentrations were identical in both groups. The duodenal ulcer patients had normal insulin responses to the intravenous glucose challenge, but the serum IR–GIP concentrations were significantly elevated in response to a 50 g oral load. Lauritsen and Moody also reported a significantly positive correlation between the integrated IR–GIP response and the integrated glucose response following oral glucose ingestion in ulcer patients but not in controls. They suggested that the increased glucose and IR–GIP concentrations in response to an oral glucose challenge were probably not caused by a reduction in glucose disposal because tolerance to intravenous glucose was normal in duodenal ulcer patients. The most likely cause of the high IR–GIP response and elevated glucose level was considered to be an increased rate of glucose absorption. The duodenal ulcer patients could be divided into two groups on the basis of their response to oral ingestion of carbohydrates: those with slightly elevated glucose and IR–GIP concentrations and those with markedly elevated concentrations. The significance of this observation was not established.

Similar findings were observed by Arnold et al. (1978) when IR–GIP release in response to a test meal was investigated in 41 patients with duodenal ulcer.

Both fasting- and meal-stimulated IR–GIP release were found to be elevated in the duodenal ulcer group. Serum IRI also increased to a significantly greater level in the patients. A relationship between the increase in serum glucose level and the serum IRI and IR–GIP was also reported. A pathological glucose tolerance was observed in 15 out of the 41 controls. The test meal induced rapid rises in IR–GIP, IRI and glucose in both groups, and the levels remained elevated for more than 5 h (Fig. 31). Subgrouping the duodenal ulcer patients into normal and pathological glucose response to the test meal revealed a relationship between glucose tolerance and elevated IR–GIP and IRI. The integrated IR–GIP, IRI and glucose responses in the duodenal ulcer patients were significantly greater than in controls. When the ulcer patient group was subdivided, the patients with pathological glucose tolerance showed significantly greater IRI and IR–GIP levels than the subgroup with normal glucose tolerance.

Creutzfeldt et al. (1978) also concluded that the observations were indicative of an enhanced glucose absorption probably occurring because of an increased rate of gastric emptying. The speed with which the elevated glucose levels were controlled indicated that no insulin resistance was present.

3. Vagotomy and Pyloroplasty

Serum glucose, IR–GIP and IRI responses have been compared by Thomford et al. (1974) in normal subjects and patients with truncal vagotomy and pyloroplasty (V and P) following ingestion of 75 g glucose. This particular study was undertaken because of the frequent problems in glucose metabolism associated with operations for peptic ulcer. In particular, Breuer et al. (1972) had demonstrated postprandial hyperglycaemia when gastric emptying effects were avoided by direct intrajejunal administration of glucose.

The fasting glucose, IR–GIP and IRI levels of the V and P patients were within the normal range. The early mean serum glucose response to ingestion of the 75 g glucose load was significantly higher in the V and P patients, but subsequently declined to significantly below fasting levels. The IRI response was faster in onset and exaggerated in the V and P patients as compared with normals. They concluded that the postprandial hypoglycaemia recognized clinically as the 'late phase' of the dumping syndrome could be associated with the augmented secretion of IR–GIP.

IV. Obesity

Karam et al. (1963) showed that obesity was characterized by a basal hyperinsulinaemia and an exaggerated IRI response to glucose. Resistance to both endogenous and exogenous insulin has been described (Rabinowitz and Zierler 1962; Perley and Kipnis 1966). The insulin resistance of obese subjects is thought to be due to the down regulation of insulin receptors produced because of the continued hyperinsulinaemia. The origin of the hyperinsulinaemia itself, however, is not understood, although Grey and Kipnis (1971) have hypothesized that excessive nutrient intake could result in islet cell hyperplasia with subsequent hyperinsulinaemia. An exaggerated enteroinsular axis

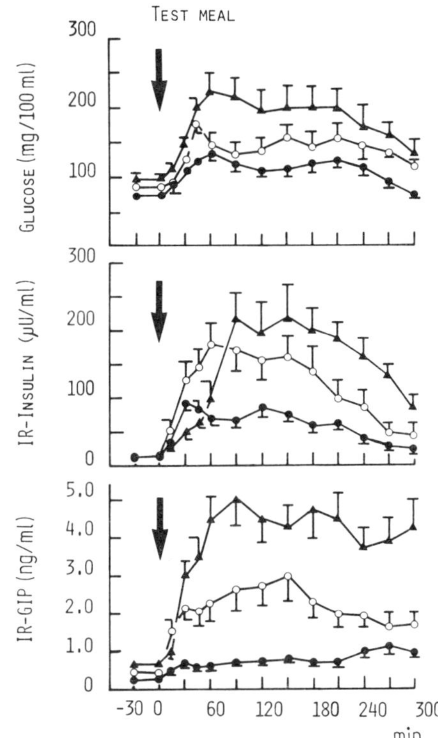

Fig. 32. Increase in serum IR-GIP, IRI and glucose levels in 15 normals (●), 31 obese subjects with normal glucose tolerance (○) and 44 obese subjects with pathological glucose tolerance (△)

could also be responsible for overproduction of insulin. Creutzfeldt et al. (1978) investigated release of IR–GIP in response to ingestion of glucose, fat and a mixed standard meal in normal weight and obese individuals (more than 30% overweight). The obese subjects were subdivided into two groups according to their glucose tolerance in response to an oral glucose tolerance test or the test meal. The oral glucose tolerance test result was considered normal (nOGTT) if the sum of the serum glucose levels at 60 and 120 min following ingestion of glucose or the test meal was less than 290 mg/dl. The subjects were considered to have a pathological OGTT result (pOGTT) if the sum of the two values was greater than 310 mg/dl. Serum IRI and IR–GIP levels were observed to be significantly elevated in obese subjects following ingestion of a high caloric test meal (Fig. 32). The responses were even more exaggerated in obese subjects with pathological glucose tolerance than in obese subjects with normal OGT. When glucose alone, 100 g in 300 ml water, was used as the provocative challenge for IR–GIP release, both serum IR–GIP and IRI levels were significantly different only in the obese group with glucose intolerance. The same IR–GIP response was seen following ingestion of 100 g triglyceride: the obese subjects released significantly greater amounts of IR–GIP than the normals, but no difference could be detected between nOGTT subjects and pOGTT subjects. The data presented suggested that ingestion of glucose was not a powerful enough provocative stimulus for IR–GIP release to demonstrate hypersecretion in obesity when normal glucose tolerance was present. This was only achieved by ingestion of a very high caloric test meal. A causal relationship between the exaggerated IRI response to the test meal and the IR–GIP response was suggested. The elevated fasting serum IRI levels and augmented insulin secretion in response to intravenous glucose injections in obesity could be considered secondary phenomena, while the primary effect is the one which brings about the high serum IRI levels. Increased postprandial IR–GIP secretion could, for example, bring about excessive beta cell stimulation with induction of islet cell hyperplasia, a further increase in IRI secretion and then a decrease in the number of insulin receptors and insulin resistance. Increased insulin resistance would lead to hyperglycaemia and further exaggerated insulin release, compounding the tissue resistance to insulin.

If the primary problem is one of exaggerated response of the GIP-producing cells, the mechanism whereby this effect occurs is not known. Suggestions are (a) that GIP hyperplasia occurs, triggered by excessive caloric intake (b) that the GIP-producing cells respond excessively and (c) that there is a breakdown in normal feedback inhibition.

The results of Creutzfeldt et al. (1978), which showed that less IR–GIP was released after the test meal than after the triglyceride load in normal subjects, whereas in the obese (pOGTT) subjects the test meal produced the better response, indicated that in the obese group a feedback control of IR–GIP by insulin or some other factor was not operating. The combination of oral glucose and oral triglyceride produced in both control and obese subjects a significantly larger IRI response than glucose alone. In the normal controls a lowered IR–GIP response occurred (intact feedback inhibition), whereas in the obese patients no change in the IR–GIP response occurred (defective feedback control). It was stated that since only the serum IRI levels were different between the normal and obese subjects after the combined oral load, the negative feedback control could not be mediated via hyperglycaemia but regulation by insulin was a possibility.

Willms et al. (1978) investigated the effect of prolonged starvation or reduced caloric intake on IR–GIP release in obese patients. Basal serum levels of IR–GIP increased within 72 h and remained elevated for the duration of the fast (21 days). The serum IRI levels, on the other hand, behaved reciprocally. The exaggerated IR–GIP and IRI responses to a test meal were reduced after 21 days starvation followed by 3 days refeeding, and the integrated values were also significantly reduced. Similar results were obtained when the provocative challenge for IR–GIP release was a 100 g load of glucose, and the IR–GIP response was lower when 100 g triglyceride was ingested.

Simple reduction of food intake to 800 cal per day (42% carbohydrate, 16% protein, 42% fat) for 21 days, followed by a 1 800–cal diet for 4 days prior to the test meal, revealed a significantly diminished IR–GIP response. The IRI response was also smaller, and the initial IRI peak occurred much earlier than before food restriction.

Willms et al. (1978) suggested that the GIP cells of the gastrointestinal tract can adapt to dietary changes, although they considered that a decreased number of GIP cells was unlikely because of the rapid reversal which can occur in a few days. Since little is known about the maintenance or 'turnover' of the GIP cell population in the gastro-intestinal tract, it is premature to exclude this probability.

Evidence has been presented suggesting that excessive nutrient ingestion promotes islet cell hyperplasia and hyperinsulinaemia and that in association with these events, there is also an exaggerated release of IR–GIP. Willms et al. (1978) concluded that the apparent association is not necessarily a causal one, but the result of several factors acting in a complex system. However, the hypersecretion of IR–GIP in obese subjects could be reversed by fasting or food restriction, and this effect was accompanied by a decreased insulin response, supporting the hypothesis that GIP is important in the pathogenesis of the hyperinsulinaemia of obesity.

V. Uraemia

The commonest abnormality of carbohydrate metabolism which is associated with uraemia is a substantially delayed utilization of a glucose load, irrespective of whether it is administered via the oral or intravenous route. There is a reduced glucose decay constant, a slightly lowered early peak of IRI release and a markedly elevated plasma IRI level during the second, prolonged phase of IRI release. In renal failure it has been concluded that two disturbances of carbohydrate metabolism exist: a reduced early insulin release and a tissue resistance towards insulin action, resulting in an underutilization of glucose.

The kidney is an important site for the clearance of several polypeptides, and O'Dorisio et al. (1977) investigated the clearance of GIP from the circulation of patients with uraemia who were undergoing chronic haemodialysis. They observed that the mean fasting serum IR–GIP levels were considerably elevated in the uraemic patients (1002 ± 145 pg/ml, mean ± SE) as compared with normal control subjects (132 ± 31 pg/ml). Haemodialysis did not significantly lower the fasting IR–GIP concentrations in the uraemic patients. Following ingestion of a test meal, mean serum IR–GIP concentrations were greater in the uraemic patients than in the control subjects between times 0 and 180 min and at 180 min; in contrast to controls, in whom the serum IR–GIP levels fell to near fasting levels at time 180 min, those of the uraemic group remained significantly elevated above basal throughout the whole of the experimental period. Using anaesthetized dogs in which the renal artery and renal vein were catheterized, O'Dorisio et al. (1977) were able to measure simultaneous renal arterial and venous serum IR–GIP concentrations during the intraduodenal perfusion of glucose. At time zero, when the serum IR–GIP level was less than 500 pg/ml, the renal A–V IR–GIP difference was 10% or less. When glucose was introduced into the duodenum and the mean arterial serum IR–GIP concentration was greater than 500 pg/ml, the renal A–V difference increased to 39% ± 11% by 30 min. A good correlation between renal arterial serum IR–GIP and renal A–V IR–GIP differences was demonstrated. The removal of IR–GIP from the circulation was partially dependent upon the arterial serum concentration. IR–GIP could not be measured in the urine. O'Dorisio et al. concluded that IR–GIP was either cleared from the circulation or modified to a non-immunoreactive form which might also render it biologically inactive. The extraction mechanism seemed to be impaired in renal failure because both basal and postprandial serum concentrations were significantly greater in the uraemic patients than in the normals. It was considered more probable that the kidney modified IR–GIP rather than simply extracting it, because haemodialysis did not remove significant amounts from the circulation and only insignificant amounts could be measured in the urine.

Kramer et al. (1977) compared plasma concentrations of IR–GIP in normal controls and uraemic patients and observed that concentrations in the latter were approximately eight times higher than in the controls. They also reported that the plasma IR–GIP levels further increased following haemodialysis. They were, however, able to detect IR–GIP in the ultrafiltrate, but at levels much lower than in plasma, and considered that the binding of IR–GIP to plasma protein accounted for this.

Serum IR–GIP levels before and after ingestion of a 100-g glucose load were measured by Matthaei et al. (1977) in normal volunteers, unselected patients on haemodialysis treatment and patients with different degrees of compensated renal failure. Serum fasting levels of IR–GIP were elevated nearly seven times above normal in patients on haemodialysis treatment and fourfold in those with compensated renal failure. After stimulation by glucose ingestion the IR–GIP levels reached a much higher level in the uraemic patients and, as described by O'Dorisio et al. (1977), the decrease in serum levels was delayed – they had not returned to prestimulation levels even 5 h after glucose ingestion. IRI levels showed a delayed increase, reaching a peak level some 2 h after glucose ingestion; the fall to basal levels, as in the case of IR–GIP levels, was slower. Matthaei et al. (1977) concluded that the enhanced insulin response to glucose may occur because IR–GIP is retained in uraemia.

VI. Other Clinical Situations

1. Pregnancy

A decrease in the 'incretin effect' by 65 % has been described by Hornnes et al. (1978) in pregnancy as compared with the same group of subjects postpartum. They subsequently investigated the release of IR–GIP in response to glucose ingestion to observe whether the decreased incretin response in late normal pregnancy could be explained by an altered IR–GIP response (Hornnes et al. 1979). Basal levels of IR–GIP were found to be unchanged in pregnancy, but IR–GIP response to a 50-g oral load of glucose was significantly diminished. It was observed that glucose absorption was at most only slightly delayed in pregnancy, indicating that changes in gastric-emptying patterns were not a cause of the altered IR–GIP response. The decrease in IR–GIP release was only moderate, whereas the earlier studies of Hornnes et al. (1978) had revealed a reduction in the incretin effect of approximately 65 %. They contended that it was possible that factors or hormones other than GIP might be involved in the mediation of the incretin effect.

2. Insulinoma

Go et al. (1979) stated that if insulin could be shown to suppress the plasma levels of glucagon and GIP directly, then its relative and absolute deficiency in diabetes could be the explanation for the elevated levels of the two hormones. They assessed the role of hyperinsulinaemia in the suppression of IR–GIP and glucagon secretion in patients with insulinoma. During the fed state, cyclical changes in glucose, IRI and IR–GIP, but not glucagon, were observed. IR–GIP responses were similar in both normals and patients with insulinoma, although mean insulin and glucagon responses were higher in the latter. In the fasted state, hypoglycaemia, hyperinsulinaemia and hyperglucagonaemia, but normal IR–GIP levels, were observed in all insulinoma patients. It was considered that the results in the patients with insulinoma supported the hypothesis that insulin did not suppress the IR–GIP response to a mixed meal and that the postulated negative-feedback relationship of insulin to IR–GIP secretion occurred only during hyperglycaemia.

G. Summary and Conclusions

I. Physiological Role

1. Gastrointestinal Actions

The initial studies which led to the isolation of GIP were based on the search for an inhibitor of gastric acid secretion using dogs with preganglionically parasympathetically denervated pouches of the body of the stomach from which gastric secretion could be collected (Brown et al. 1969; Brown and Pederson 1970). The search for the inhibitor in preparations containing CCK–PZ was prompted by the physiological studies of Brown and Pederson (1970), in which evidence was presented that CCK–PZ itself could not be the inhibitor of gastric acid secretion in the dog.

A polypeptide was eventually isolated from a CCK–PZ containing extract, and an amino acid sequence was reported (Brown et al. 1969; Brown et al. 1970; Brown 1971; Brown and Dryburgh, 1971). A recent correction has been made to this sequence (Jörnvall et al. 1981), and the presence of a minor contaminant in purest GIP preparations, having the sequence GIP_{3-42}, confirmed.

The heterogeneous GIP (with the structurally similar minor contaminant) was shown to be a powerful inhibitor of acid secretion in dogs with denervated gastric pouches (Pederson and Brown 1972) when secretion was stimulated by pentagastrin and to be less effective as an inhibitor when histamine was used. In the innervated stomach of man and dog, GIP was shown to be a less effective inhibitory agent (Maxwell et al. 1980; Soon–Shiong et al. 1979a). A partial explanation of this lack of inhibitory effect in the innervated stomach has been presented. Soon–Shiong et al. (1979b) demonstrated that intravenous administration of a stable parasympathomimetic agent would reverse the inhibitory action of GIP on acid secretion in dogs with denervated pouches. In addition, McIntosh et al. (1979) and Brown et al. (1980) were able to stimulate SLI release from the isolated perfused stomach of the rat with 5.0 and 50 ng/ml GIP. They suggested that GIP may exert its inhibitory effect on acid secretion via release of somatostatin. The role of parasympathetic innervation in inhibiting the release of SLI by GIP was also demonstrated by these workers. They were able to inhibit SLI release by GIP by introducing acetylcholine into the perfusate and also by electrical stimulation of the distal end of the sectioned vagal nerves. It must be concluded that the inhibitory effect of GIP on the parietal cell is not a direct one, nor is it by any means simple. In the physiological situation where GIP is released in response to a meal, numerous interacting mechanisms are activated, both stimulatory and inhibitory, nervous and hormonal, gastric and duodenal, active and passive. To dismiss GIP as an important component in the regulation of gastric secretion because of its poor effect in the presence of parasympathetic innervation is premature. Worthy of consideration are the possible interactions of GIP with other inhibitory mechanisms of duodenal origin.

2. Metabolic Actions

The study by Dupré et al. (1973), in which the insulinotropic action of GIP was demonstrated in man, concluded some 50 years of searching for the gastro-intestinal insulin-releasing factor. The glucose-dependent nature of the insulinotropic effect of GIP was also described and the physiological significance indicated when insulin release was induced with a dose of GIP which elevated the serum IR–GIP levels to approximately 1.0 ng/ml. A threshold glucose concentration for the insulinotropic effect of GIP in the isolated perfused rat pancreas was shown by Pederson and Brown (1975) to be 5.5 m*M*, a figure which also applies to man. The magnitude of the insulinotropic effect is also glucose concentration-dependent, and in the rat the greatest potentiation occurred at approximately 16 m*M*, when an approximately five-fold increase in insulin release with GIP was observed. Insulin release in the presence of GIP exceeded by approximately 400% the maximum achievable with glucose alone.

Pederson and Brown (1978) showed a relationship between arginine, GIP and insulin release, and their data were suggestive of a common mechanism of action for arginine and GIP in the presence of glucose. The glucagonotropic effect of GIP in the isolated perfused rat pancreas has not, as yet, been demonstrated in man.

The action of GIP on rat isolated pancreatic islets was shown to be quite different from that on the isolated perfused pancreas and much weaker (Schauder et al. 1975). A concentration of GIP of 10 μg/ml was required to initiate the response, which was different from the perfused pancreas response in that the maximum response to glucose could not be increased by GIP. Brown et al. (1980) proposed several possible explanations the poor response from the isolated islets:

1. The islets were isolated following digestion with collagenase, which may have destroyed the receptors for GIP.
2. A static system for islet incubation was used; this could allow the accumulation of substances which could possibly interfere with insulin release, e.g. somatostatin or metabolites.
3. In the isolated perfused pancreas, delivery of glucose etc. to the cells of the islets is achieved via the circulation and microcirculation, whereas during incubation of the isolated islets access to the cells can only be achieved by diffusion.
4. The action of GIP on insulin release could be an indirect one, acting at a site other than on the B cell or other islet cell, and this site would not be present following islet isolation.

Brown et al. (1980) also brought attention to the biphasic nature of insulin release in response to GIP. A possible explanation for this phenomenon was suggested. GIP was presented to the isolated perfused pancreas as a square-wave stimulus and the sudden large increase in the stimulus (0 to 1.0 ng/ml) may have been of sufficient magnitude to evoke a spike release of somatostatin, thereby producing a transient inhibition of insulin release. Ipp et al. (1977) were able to release somatostatin in the isolated perfused dog pancreas (threefold increase) with pharmacological amounts of GIP (58 ng/ml) in the presence of 100 mg/dl glucose.

Little is known about the mechanism of action of GIP.

II. Pathophysiological Role

Gastric inhibitory polypeptide has not been causally implicated in a specific disease state although changes in blood levels have been described for many (Cataland and O'Dorisio 1980; Creutzfeldt and Ebert 1977). Many controversies have arisen and will continue to do so because of poor characterization of antisera used in radioimmunoassays. Brown et al. (1979) drew attention to the heterogeneity of IR–GIP and the different measurements which could be obtained on the same serum samples using two antibodies, both raised in guinea pigs, by the same immunization procedures. They further indicated that the ratio of the heterogeneous forms of IR–GIP followed no distinct pattern when serum samples were examined following ingestion of glucose or triglyceride.

Apparent hypersecretion of GIP has been shown to occur in situations where hyperinsulinaemia is found, e.g. in obesity and maturity-onset diabetes, and hyposecretion was shown to occur in coeliac disease. Questions requiring an answer include: can a disturbed carbohydrate metabolism be induced by chronic administration of exogenous GIP, and how can GIP release be inhibited? Answers to these questions are essential before the true implications of this hormone in clinical endocrinology can be assessed.

References

Alberti KGMM, Christensen SE, Iversen J, Seyer-Hansen K, Christensen NJ, Prange-Hansen A, Lundbaek K, Ovstov H (1973) Inhibition of insulin secretion by somatostatin. Lancet II:1299–1301

Andersen DK, Elahi D, Brown JC, Tobin JD, Andres R (1978) Oral glucose augmentation of insulin secretion: Interactions of gastric inhibitory polypeptide with ambient glucose and insulin levels. J Clin Invest 62:152–161

Andersen DK, Putnam WS, Hanks JB, Wise JE, Lebovitz HE, Jones RS (1980) Gastric inhibitory polypeptide (GIP) suppression of hepatic glucose production. Regulat Pept 1:4

Andersson S (1960a) Inhibitory effects of acid in antrum-duodenum on fasting gastric secretion in Pavlov and Heidenhain pouch dogs. Acta Physiol Scand 49:42–56

Andersson S (1960b) Inhibitory effects of hydrochloric acid in antrum and duodenum on gastric secretory responses to test meal in Pavlov and Heidenhain pouch dogs. Acta Physiol Scand 49:231–241

Andersson S (1960c) Inhibitory effects of hydrochloric acid in antrum and duodenum on histamine stimulated gastric secretion in Pavlov and Heidenhain pouch dogs. Acta Physiol Scand 50:186–196

Andersson S, Grossman MI (1965) Effect of denervation and subsequent resection of antral pouches on secretion from Heidenhain pouches in response to gastrin and histamine. Gastroenterology 51:4–9

Andersson S, Nilsson G, Uvnas B (1965) Inhibition of gastric secretion by acid in proximal and distal duodenal pouches. Acta Physiol Scand 65:191–192

Arnold R, Creutzfeldt W, Ebert R, Becker HD, Borger HW, Schafmayer A (1978) Serum gastric inhibitory polypeptide (GIP) in duodenal ulcer disease: Relationship to glucose tolerance, insulin, and gastrin release. Scand J Gastroenterol 13:41–47

Babkin BP (1944) Secretory mechanisms of the digestive glands, 2nd edn. Hoeber, New York, p 642

Bank S, Marks IN, Moshal NG, Efron G, Silber R (1963) Pancreatic function test, method and normal values. S Afr Med J 37:1061–1066

Bank S, Jackson WPU, Keller P, Marks IN (1968) Serum insulin response to glucose in pancreatic diabetes. Postgrad Med J 44:214–217

Barbezat GO, Grossman MI (1971) Intestinal secretion: Stimulation by peptides. Science 174:422–424

Bataille D, Freychet P, Rosselin G (1974) Interactions of glucagon, gut glucagon, vasoactive intestinal polypeptide and secretin with liver and fat cell membranes: Binding to specific sites and stimulation of adenylate cyclase. Endocrinology 95:713–721

Baumert JE, Cataland S, Tetirick CE, Pace WG, Mazzaferri EL (1978) Effect of atropine on meat-stimulated gastrin and gastric inhibitory polypeptide (GIP) release. J Clin Endocrinol Metab 46:473–476

Becker HD, Reeder DD, Thompson JC (1973) Effect of glucagon on circulating gastrin. Gastroenterology 65:28–35

Besterman HS, Sarson DL, Johnston DI, Stewart JS, Guerin S, Bloom SR, Blackburn AM, Patel HR, Modigliani R, Mallison CH (1978) Gut–hormone profile in coeliac disease. Lancet 15 April, 785–788

Beyreiss K, Muller F, Strack E (1964) Über die Resorption von Monosacchamden I. Der Einfluß von Insulin auf die Resorption der Galaktose. Z Gesamte Exp Med 138:277–288

Bloom SR (1974) Hormones of the gastrointestinal tract. Br Med Bull 30:62–67

Bloom SR, Bryant MG, Cochrane UPS (1975) Normal distribution and post prandial release of gut hormones. Clin Sci Mol Med 49:3

Bodansky M, Ondetti MA, Levine SD (1966) Synthesis of a heptacosapeptide amide with the hormonal activity of secretin. Chem Ind 42:1757–1758

Boden G, Essa N, Owen OE (1975) Effects of intraduodenal amino acids, fatty acids and sugars on secretin concentrations. Gastroenterology 68:722–727

Botha JL, Vinik AI, Brown JC (1976) Gastric inhibitory polypeptide in chronic pancreatitis. J Clin Endocrinol Metab 42:791–797

Botha JL, Vinik AI, Child PT (1978) Gastric inhibitory polypeptide in acquired pancreatic diabetes: Effects of insulin treatment. J Clin Endocrinol Metab 47:543–549

Bottger I, Faloona GR, Unger RH (1972) Response of islet cell and gut hormones to fat absorption: An "enteroinsular axis" for fat. Clin Res 20:542

Bottger I, Dobbs R, Faloona GR, Unger RH (1973) The effects of triglyceride absorption upon glucagon, insulin and gut glucagon-like immunoreactivity. J Clin Invest 52:2532–2541

Breuer RI, Moses H, Hagen TC, Zuckerman L (1972) Gastric operations and glucose homeostasis. Gastroenterology 62:1109–1119

Brown JC (1971) A gastric inhibitory polypeptide. I. The amino acid composition and the tryptic peptides. Can J Biochem 49:255–261

Brown JC (1974) Gastric inhibitory polypeptide (GIP). In: Taylor S (ed) Endocrinology 1973. Heinemann, London, pp 276–284

Brown JC, Dryburgh JR (1971) A gastric inhibitory polypeptide II: The complete amino acid sequence. Can J Biochem 49:867–872

Brown JC, Dryburgh JR (1979) Gastric inhibitory polypeptide. In: Jaffe B (ed) Methods of hormone radioimmunoassay, 2nd edn. Academic, New York London, pp 541–552

Brown JC, Magee DF (1967) Inhibitory action of cholecystokinin on acid secretion from Heidenhain pouches, induced by endogenous gastrin. Gut 8:29–31

Brown JC, Otte SC (1978) Gastrointestinal hormones and control of insulin secretion. Diabetes 27:782–789

Brown JC, Otte SC (1979a) GIP and the enteroinsular axis. Endocrinol Metab 8/2: 365–377

Brown JC, Otte SC (1979b) Clinical studies with gastric inhibitory polypeptide. World J Surg 3:553–558

Brown JC, Pederson RA (1970) A multiparameter study on the action of preparations containing cholecystokinin–pancreozymin. Scand J Gastroenterol 5:537–541

Brown JC, Johnson LP, Magee DF (1967) The inhibition of induced motor activity in transplanted fundic pouches. J Physiol (Lond) 188:45–52

Brown JC, Pederson RA, Jorpes E, Mutt V (1969) Preparation of highly active enterogastrone. Can J Physiol Pharmacol 47:113–114

Brown JC, Mutt V, Pederson RA (1970) Further purification of a polypeptide demonstrating enterogastrone activity. J Physiol 209:57–64

Brown JC, Dryburgh JR, Moccia P, Pederson RA (1975a) The current status of GIP. In: Thompson JC (ed) Gastrointestinal hormones. University of Texas Press, Austin, pp 537–547

Brown JC, Dryburgh JR, Ross SE, Dupre J (1975b) Identification and actions of gastric inhibitory polypeptide. Recent Prog Horm Res 31:487–532

Brown JC, Dryburgh JR, Frost JL, Otte SC, Pederson RA (1978) Properties and actions of GIP. In: Bloom SR (ed) Gut hormones. Churchill Livingstone, Edinburgh, pp 277–282

Brown JC, McIntosh CHS, Muller M, Otte SC, Pederson RA (1979) GIP and the enteroinsular axis. In: Miyoshi A (ed) Gut peptides. Kodansha, Tokyo Elsevier/North Holland Biomedical Press, Amsterdam New York Oxford, pp 162–168

Brown JC, Koop H, McIntosh CHS, Otte SC, Pederson RA (1980) Physiology of gastric inhibitory polypeptide. Front Horm Res 7:132–144

Brown JC, Dahl M, Kwauk S, McIntosh CHS, Muller M, Otte SC, Pederson RA (1981) Properties and actions of GIP. In: Bloom SR, Polak JM (eds) Gut hormones, 2dn edn. Churchill Livingstone, Edinburgh, pp 248–255

Buchanan KD (1973) Studies on the pancreatic enteric hormones. PhD thesis, Queens University of Belfast

Buchanan KD, McKiddle MT, Lindsay AC, Manderson WG (1967) Carbohydrate metabolism in duodenal ulcer patients. Gut 8:325–331

Buchanan KD, Vance JE, Morgan A, Williams RH (1968) Effect of pancreozymin on insulin and glucagon levels in blood and bile. Am J Physiol 215:1293–1298

Buchanan KD, Vance JE, Williams RH (1969) Insulin and glucagon release from isolated islets of Langerhans. Effect of enteric factors. Diabetes 18:381–386

Buffa R, Polak JM, Pearse AGE, Solcia E, Grimelius L, Capella (1975) Identification of the intestinal cell storing gastric inhibitory peptide. Histochemistry 43:249–255

Burhol PG, Jorde R, Waldun HL (1980) Radioimmunoassay of plasma gastric inhibitory polypeptide (GIP), release of GIP after a test meal and duodenal infusion of bile, and immunoreactive plasma GIP components in man. Digestion 20:336–345

Camble R, Cotton R, Dutta AS, Gormley JJ, Hayward CF, Morley JC, Smither MI (1973) On the synthesis of a porcine gastric inhibitory peptide. In: Hanson H, Jakubke HD (eds) Peptide 1972. North Holland, Amsterdam, pp 200–209

Carpenter CCJ, Sack RB, Feeley JC, Steenberg RW (1968) Site and characteristics of electrolyte loss and effect of intraluminal glucose in experimental canine cholera. J Clin Invest 47:1210–1220

Cataland S, O'Dorisio TM (1980) Pathology of gastric inhibitory polypeptide. Front Horm Res 7:145–154

Cataland S, Crockett SE, Brown JC, Mazzaferri EL (1974) Gastric inhibitory polypeptide (GIP) stimulation by oral glucose in man. J Clin Endocrinol Metab 39:223–228

Cataland S, O'Dorisio TM, Brooks R, Mekhjian HS (1977) Stimulation of gastric inhibitory polypeptide in normal and duodenal ulcer patients. Gastroenterology 73:19–22

Cataland S, O'Dorisio TM, Crockett SE, Mekhjian HS (1978) Gastric inhibitory polypeptide (GIP) and insulin release in duodenal ulcer patients. J Clin Endocrinol Metab 47:615–619

Cerasi E, Efendic S, Luft R (1973) Dose-response relation between plasma-insulin and blood-glucose levels during oral glucose loads in prediabetic and diabetic subjects. Lancet I: 794–797

Chey WY, Rhodes RA, Lee KY, Hendricks J (1975) Radioimmunoassay of secretin: Further studies. In: Thompson JC (ed) Gastrointestinal hormones. University of Texas Press, Austin, pp 269–281

Cleator IGM, Gourlay RH (1975) Release of immunoreactive gastric inhibitory polypeptide (IR–GIP) by oral ingestion of food substances. Am J Surg 130:128–135

Code DF, Watkinson G (1955) Importance of vagal innervation in the regulatory effect of acid in the duodenum on gastric secretion of acid. J Physiol (Lond) 130:233–252

Crane RK (1955) Na^+-dependent transport in the intestine and other animal tissues. Fed Proc 24:1000–1006

Crane RK (1968) Absorption of sugars, vol III. Washington, American Physiology Society, pp 1323–1358

Creutzfeldt W (1979) The incretin concept today. Diabetologia 16:75–85

Creutzfeldt W, Ebert R (1977) Release of gastric inhibitory polypeptide (GIP) to a test meal under normal and pathological conditions in man. In: Bajaj JS (ed) Diabetes. Excerpta Medica, Amsterdam, pp 63–75

Creutzfeldt W, Feurle G, Ketterer H (1970) Effect of gastrointestinal hormones on insulin and glucagon secretion. N Engl J Med 282:1139–1141

Crockett SE, Cataland S, Falko J, Mazzaferri EL (1975) Gastric inhibitory polypeptide: Responses to variable doses of glucose in normal subjects and abnormal responses to oral glucose in patients with adult onset diabetes mellitus. Diabetes 24:413

Creutzfeldt W, Ebert R, Arnold R, Frerichs H, Brown JC (1976) Gastric inhibitory polypeptide (GIP), gastrin and insulin: Response to test meal in coeliac disease and after duodeno-pancreatectomy. Diabetologia 12:279–286

Creutzfeldt W, Ebert R, Willms B, Frerichs H, Brown JC (1978) Gastric inhibitory polypeptide (GIP) and insulin in obesity: Increased response to stimulation and defective feedback control serum levels. Diabetologia 14:15–24

Creutzfeldt W, Ebert R, Caspary W, Folsch U, Lembcke B (1979) Dependence of GIP release upon absorption of nutrients. In: Miyoshi A (ed) Gut peptides. Kodansha, Tokyo, Elsevier/North Holland, Amsterdam, pp 79–87

Creutzfeldt W, Talaulicar M, Ebert R, Willms B (1980) Inhibition of gastric inhibitory polypeptide (GIP) release by insulin and glucose in juvenile diabetes. Diabetes 29:140–145

Crockett SE, Cataland S, Falko JM, Mazzaferri EL (1976a) The insulinotropic effect of endogenous gastric inhibitory polypeptide in normal subjects. J Clin Endocrinol Metab 42:1098–1103

Crockett SE, Mazzaferri EL, Cataland S (1976b) Gastric inhibitory polypeptide (GIP) in maturity-onset diabetes mellitus. Diabetes 25:931–935

Day JJ, Komarov SA (1939) Glucose and gastric secretion. Am J Dig Dis 6:169–175

Day JJ, Webster DR (1935) The autoregulation of the gastric secretion. Am J Dig Dis Nutr 2: 527–531

Denniss AR, Young JA (1978) Modification of salivary duct electrolyte transport in rat and rabbit by physalaemin, VIP, GIP and other enterohormones. Pfluegers Arch 376:73–80

Desbuquois B, Aurbach GD (1971) Use of polyethylene glycol to seperate free and antibody bound peptide hormones in radioimmunoassays. J Clin Endocrinol 33:732–738

Dixon W, Wadia JH (1926) The action of intestinal extracts. Br Med J I:820

Dryburgh JR (1977) Immunological techniques in the investigation of the physiological functions of gastric inhibitory polypeptide and motilin. PhD Thesis, University of British Columbia

Dryburgh JR, Hampton SM, Morgan LM, Marks V (1980) C-peptide-induced inhibition of GIP release. Regulat Pept [Supp] 1:28

Dupré J (1964) An intestinal hormone affecting glucose disposal in man. Lancet II: 672–673

Dupré J, Rojas L, White JJ, Unger RH, Beck JC (1966) Effects of secretin on insulin and glucagon in portal and peripheral blood in man. Lancet II:26–27

Dupré J, Curtis JD, Unger RH, Waddell RW, Beck JC (1969) Effects of secretin, pancreozymin or gastrin on the response of the endocrine pancreas to administration of glucose or arginine in man. J Clin Invest 48:748–757

Dupré J, Ross SA, Watson D, Brown JC (1973) Stimulation of insulin secretion by gastric inhibitory polypeptide in man. J Clin Endocrinol Metab 37:826–828

Dupré J, Greenidge N, McDonald TJ, Ross SA, Rubinstein D (1976) Inhibition of actions of glucagon in adipocytes by gastric inhibitory polypeptide. Metabolism 25:1197–1199

Ebert R, Brown JC (1976) Effect of gastric inhibitory polypeptide (GIP) on lipolysis and cyclic AMP levels in isolated fat cells. Eur J Clin Invest 6:327

Ebert R, Creutzfeldt W (1978) Aspects of GIP pathology. In: Bloom SR (ed) Gut hormones. Churchill Livingstone, Edinburgh, pp 294–300

Ebert R, Creutzfeldt W (1980) Decreased GIP secretion through impairment of absorption. Front Horm Res 7:192–201

Ebert R, Creutzfeldt W, Brown JC, Frerichs H, Arnold R (1976a) Response of gastric inhibitory polypeptide (GIP) to test meal in chronic pancreatitis–relationship to endocrine and exocrine insufficiency. Diabetologia 12:609–612

Ebert R, Frerichs H, Creutzfeldt W (1976b) Serum gastric inhibitory polypeptide (GIP) response in patients with maturity onset diabetes and in juvenile diabetes. Diabetologia 12:388

Ebert R, Arnold R, Creutzfeldt W (1977) Lowering of fasting and food stimulated serum immunoreactive gastric inhibitory polypeptide (GIP) by glucagon. Gut 8:121–127

Ebert R, Illmer K, Creutzfeldt W (1978) Abolishment of the incretin effect in rats by infusion of gastric inhibitory polypeptide (GIP) antibodies. Scand J Gastroenterol 13:53

Ebert R, Illmer K, Creutzfeldt W (1979) Release of gastric inhibitory polypeptide (GIP) by intraduodenal acidification in rats and humans and abolishment of the incretin effect of acid by GIP-antiserum in rats. Gastroenterology 76:515–523

Eckel RH, Fujimoto WY, Brunzell JD (1979) Gastric inhibitory polypeptide enhanced lipoprotein lipase activity in cultured preadipocytes. Diabetes 28:1141–1142

Efendic S, Hokfeldt T, Luft R (1978) Somatostatin. Adv Metab Disord 9:367–424

Elahi D, Andersen DK, Brown JC, Debas H, Hershcopf RJ, Raizes GS, Tobin JD, Andres R (1979) Pancreatic α- and β-cell responses to GIP infusion in normal man. Am J Physiol 237/2: 185–191

Elrick ML, Stimmler L, Hlad CJ, Arai Y (1964) Plasma insulin response to oral and intravenous glucose administration. J Clin Endocrinol 24:1076–1082

Ewald CA, Boas J (1886) Beiträge zur Physiologie und Pathologie der Verdauung II. Virchows Arch [Pathol Anat] 104:271–305

Falko JM, Crockett SE, Cataland S, Mazzaferri EL (1975) Gastric inhibitory polypeptide (GIP) stimulated by fat ingestion in man. J Clin Endocrinol Metab 41:260–265

Fara JW, Salazar AM (1978) Gastric inhibitory polypeptide increases mesenteric blood flow. Proc Soc Exp Biol Med 158:446–448

Fara JW, Rubinstein ER, Sonnenschein RR (1972) Intestinal hormones in mesenteric vasodilation after intraduodenal agents. Am J Physiol 223:1058–1067

Farrell JI, Ivy AC (1926) Studies on the motility of the transplanted gastric pouch. Am J Physiol 46: 227–228

Feng TP, Hou HC, Lim RKS (1929) On the mechanism of the inhibition of gastric secretion by fat. Chin J Physiol 3:371–378

Fujimoto WY, Ensinck JW, Merchant W, Williams RH, Smith PH, Johnson DG (1978) Stimulation by gastric inhibitory polypeptide of insulin and glucagon secretion by rat islet cell cultures. Proc Soc Exp Biol Med 157:89–93

Füssganger K, Straub K, Goberna R, Jaros P, Schroder KE, Raptis S, Pfeiffer EF (1969) Primary secretion of insulin and secondary release of glucagon from the isolated perfused rat pancreas following stimulation with pancreozymin. Horm Metab Res 1:224–227

Gaginella TS, Mekhjian HS, O'Dorisio TM (1978) Vasoactive intestinal peptide: Quantification by radioimmunoassay in isolated cells, mucosa, and muscle of the hamster intestine. Gastroenterology 74:718

Gillespie IE, Grossman MI (1964) Inhibitory effect of secretin and cholecystokinin on Heidenhain pouch responses to gastrin extract and histamine. Gut 5:342–345

Go VLW, Reilly WM (1975) Problems encountered in the development of the cholecystokinin radioimmunoassay. Gastrointestinal hormones. Austin, University of Texas Press, pp 295–299

Go VLW, Hofmann AF, Summerskill W.H.J. (1970) Pancreozymin bioassay in man based on pancreatic enzyme secretion: Potency of specific amino acids and other digestive products. J Clin Invest 49:1558–1564

Go VLW, Nelson RL, McCullouch AJ, Verdonk CA, Service FJ (1979) Interrelationships among gastric inhibitory polypeptide, insulin and glucagon in health and insulinoma. In: Miyoshi A (ed) Gut peptides. Kodansha, Tokyo, Elsevier/North Holland, Amsterdam New York Oxford, pp 337–342

Goodfriend TL, Levine L, Fasman GD (1964) Antibodies to bradykinin and angiotensin: A use of carbodiimides in immunology. Science 144:1344–1346

Goulder TJ, Alberti KGMM, Jenkins DJA (1978) The effect of added fibre to the glucose and metabolic response to a mixed meal in normal and diabetic subjects. Diabetes Care 1: 351–355

Gray GM (1975) Carbohydrate digestion and absorption: Role of the small intestine. N Eng J Med 292:1225–1230

Gray JS, Bradley WB, Ivy AC (1937) On the preparation and biological assay of enterogastrone. Am J Physiol 118:463–476

Greengard H, Atkinson AJ, Grossman MI, Ivy AC (1946) The effectiveness of parenterally administered "enterogastrone" in the prophylaxis of recurrences of experimental and clinical peptic ulcer. Gastroenterology 7:625–649

Greenwood FC, Hunter WM, Glover JS (1963) The preparation of ^{131}I labelled human growth hormone of high specific radioactivity. Biochem J 89:114–123

Gregory RA, Tracy HJ (1964) The constitution and properties of two gastrins extracted from hog antral mucosa. Gut 5:103–107

Grey NJ, Kipnis DM (1971) Effect of diet composition on the hyperinsulinaemia of obesity. N Engl J Med 285:827–831

Hansky J, Soveny C, Korman KG (1973) The effect of glucagon on serum gastrin. I. Studies in normal subjects. Gut 14:457–458

Harvey RF, Dowsett L, Hartog M, Read AE (1973) A radioimmunoassay for cholecystokinin–pancreozymin. Lancet II:826–827

Helmon CA, Barbezat GO (1977) The effect of gastric inhibitory polypeptide on human jejunal water and electrolyte transport. Gastroenterology 72:376–379

Hornnes P, Kuhl C, Klebe JG (1978) Diminished gastrointestinal potentiation of insulin secretion in human pregnancy. Diabetologia 15:165–168

Hornnes P, Kuhl C, Lauritsen KB (1979) Diminished gastric inhibitory polypeptide response to oral glucose in late human pregnancy. J Clin Endocrinol Metab 48:506–508

Humphrey CS, Dykes JRW, Johnston D (1972) Glucose tolerance and insulin secretion in patients with chronic duodenal ulcer. Br Med J 4:393–396

Ipp E, Dobbs RE, Harris V, Arimura A, Vale W, Unger RH (1977) The effects of gastrin, GIP, secretin and the octapeptide of cholecystokinin upon immunoreactive somatostatin release by the perfused canine pancreas. J Clin Invest 60:1216–1219

Iverson J (1971) Secretion of glucagon from the isolated perfused canine pancreas. J Clin Invest 50:2123–2136

Johns EW (1967) The electrophoresis of histones in polyacrylamide gel and their quantitative determination. Biochem J 104:78–81

Johnson LP, Magee DF (1965) Cholecystokinin–pancreozymin extracts and gastric motor inhibition. Surg Gynecol Obstet 121:557–562

Johnson LR, Grossman MI (1968) Secretin: The enterogastrone released by acid in the duodenum. Am J Physiol 215:885–888

Jörnvall H, Carlquist M, Kwauk S, Otte SC, McIntosh CHS, Brown JC, Mutt V (1981) Amino acid sequence and heterogeneity of gastric inhibitory polypeptide (GIP). FEBS Lett 123:205–210

Jorpes JE, Mutt V (1961) Gastrointestinal hormones, secretin and cholecystokinin–pancreozymin. Ann Intern Med 55:395–405

Jorpes JE, Mutt V (1966) Cholecystokinin and pancreozymin–one single hormone. Acta Physiol Scand 66:196–202

Jorpes JE, Mutt V, Toczko K (1964) Further purification of cholecystokinin and pancreozymin. Acta Chem Scand 18:2408–2410

Kalk WJ, Vinik AI, Bank S, Keller P, Jackson WPU (1974) Selective loss of beta cell response to glucose in chronic pancreatitis. Horm Metab Res 6:95–98

Karam JH, Grodsky GM, Forsham PH (1963) Excessive insulin response to glucose in obese subjects as measured by immunochemical assay. Diabetes 12:197–205

Konturek S, Grossman MI (1965) Effect of perfusion of intestinal loops with acid, fat or dextrose on gastric secretion. Gastroenterology 49:481–489

Konturek SJ, Tasler J, Obtuzowicz W (1972) Localization of cholecystokinin release in the intestine of dog. Am J Physiol 222:16–20

Kosaka T, Lim RKS (1930) Demonstration of the humoral agent in fat inhibition of gastric secretion. Proc Soc Exp Biol Med 27:890–891

Kosaka T, Lim RKS, Ling SM, Liu AC (1932) On the mechanism of the inhibition of gastric secretion by fat. A gastric–inhibitory agent obtained from the intestinal mucosa. Chin J Physiol 6:107–126

Kramer P, Matthaei D, Arnold R, Ebert R, Kobberling D, McIntosh C, Schwinn E, Scheler F, Ludwig H, Reichel J, Spitella G (1977) Changes of plasma concentration and elimination of various hormones by haemofiltration. Proc Eur Dial Transplant Assoc 14:144–150

Kuhl C, Hornnes PJ, Jensen SL, Lauritsen KB (1980) Gastric inhibitory polypeptide and insulin: Response to intraduodenal and intravenous glucose infusions in fetal and neonatal pigs. Endocrinology 107:1446–1450

Kuzio M, Dryburgh JR, Malloy KM, Brown JC (1974) Radioimmunoassay for gastric inhibitory polypeptide. Gastroenterology 66:357–364

Labarre J (1932) Sur les possibilites d'un traitment du diabetes par l'incretine. Bull Acad R Med Belg 12:620–634

Labarre J, Still EV (1930) Studies on the physiology of secretin. Am J Physiol 91:649–653

Larrimer JN, Mazzaferri EL, Cataland S, Mekhjian HS (1978) Effect of atropine on glucose-stimulated gastric inhibitory polypeptide. Diabetes 27:638–642

Laughton NM, Macallum AB (1932) The relation of duodenal mucosa to the internal secretion of the pancreas. Proc R Soc Lond [Biol] 111:37–46

Lauritsen KB, Moody AJ (1978) The response of gastric inhibitory polypeptide (GIP) and insulin to glucose in duodenal ulcer patients. Diabetologia 14:149–153

Lerner RL, Porte D Jr (1970) Uniphasic insulin responses to secretin stimulation in man. J Clin Invest 49:2276–2280

Lerner RL, Porte D (1972) Studies of secretin–stimulated insulin responses in man. J Clin Invest 51:2205–2210

Ljungberg S (1962) Biologisk styrkebestamning av cholecystokinin. Sven Farm Tidskr 68:351–354

Loew ER, Gray JS, Ivy AC (1940) Is a duodenal hormone involved in carbohydrate metabolism. Am J Physiol 129:659–663

Long JF, Brooks FP (1965) Relation between inhibition of gastric secretion and absorption of fatty acids. Am J Physiol 209:447–451

Magee DF, Nakamura M (1966) Action of pancreozymin preparations on gastric secretion. Nature 212:1487–1488

Marks V, Samols E (1969) Action de differents stimuli (galactose, fructose, et lipides) sur l'insulin secretion humaine influences du tractus gastro-intestinal. J Annu Diabetol Hotel Dieu 9:179–190

Marks V, Turner DS (1977) The gastrointestinal hormones with particular reference to their role in the regulation of insulin secretion. Essays Med Biochem 3:109–152

Martin EW, Sirinek KR, Crockett SE, O'Dorisio TM, Mazzaferri EL, Thomford NR, Cataland S

(1975) Release of gastric inhibitory polypeptide: Comparison of hyperosmolar carbohydrate solutions as stimuli. Surg Forum 26:381–382

Matthaei D, Ebert R, Schauder P, Frerichs H, Creutzfeldt W, Scheler (1977) Serum levels of immunoreactive gastric inhibitory polypeptide and insulin in uraemic patients. Proc Eur Dial Transplant Assoc 14:550–556

Maxwell V, Shulkes A, Brown JC, Solomon TE, Walsh JH, Grossman MI (1980) Effect of gastric inhibitory polypeptide on pentagastrin-stimulated acid secretion in man. Dig Dis Sci 25:113–116

May JM, Williams RH (1978) The effect of endogenous gastric inhibitory polypeptide on glucose-induced insulin secretion in mild diabetes. Diabetes 27:849–855 ·

McIntosh C, Arnold R (1978) The radioimmunoassay and physiology of somatostatin in the pancreas and gastrointestinal tract. Gastroenterol 16:330

McIntosh CHS, Pederson RA, Koop H, Brown JC (1979) Inhibition of GIP stimulated somatostatine-like (SLI) by acetylcholine and vagal stimulation. In: Miyoshi A (ed) Gut peptides. Kodansha, Tokyo, Elsevier/North Holland, Amsterdam, pp 100–104

McIntyre N, Holdsworth CD, Turner DS (1964) New interpretation of oral glucose tolerance. Lancet II:20–21

Menguy R (1960) Studies on the role of pancreatic and biliary secretions in the mechanism of gastric inhibition by fat. Surgery 48:195–200

Meyer JH (1975) Release of secretin and cholecystokinin. Gastrointestinal hormones. University of Texas Press, Austin, pp 475–489

Meyer JH, Grossman MI (1972) Comparison of D- and L-phenylalanine as pancreatic stimulants. Am J Physiol 222:1058–1063

Meyer JH, Spingola LJ, Jones RS (1973) Canine pancreatic responses to intestinally perfused L-amino acids and peptides (Abstracts). Clin Res 21:207

Moody AJ (1977) Insulin releasing polypeptides of the gut. In: Bajaj JS (ed) Diabetes. Excerpta Medica, Amsterdam, pp 76–82

Moore B, Edie ES, Abram JH (1906) On the treatment of diabetes mellitus by acid extract of duodenal mucous membrane. Biochem J 1:28–38

Morgan LM (1979) Immunoassayable gastric inhibitory polypeptide: Investigations into its role in carbohydrate metabolism. Ann Clin Biochem 16:6–14

Morgan LM, Morris BA, Marks V (1978) Radioimmunoassay of gastric inhibitory polypeptide. Ann Clin Biochem 15:172–177

Morgan LM, Wright JW, Marks V (1979) The effect of oral galactose on GIP and insulin secretion in man. Diabetologia 16:235–239

Moroder L, Hallett A, Thamm P, Wilschowitz L, Brown JC, Wunsch E (1978) Studies on gastric inhibitory polypeptide: Synthesis of the octatriacontapeptide GIP^{1-38} with full insulinotropic activity. Scand J Gastroenterol [Suppl 49] 13:129

Murat JE, White TT (1966) Stimulation of gastric secretion by commercial cholecystokinin extracts. Proc Soc Exp Biol Med 123:593–594

Mutt V (1959) Preparation of highly purified secretin. Ark Kemi 15:69–74

Mutt V, Jorpes JE (1966) Secretin isolation and determination of structure. In: Back N, Marini L, Paoletti R (Ed) Plenum Press, New York, 1968 pp 569–574 Proceedings of the Fourth International Symposium on the Chemistry of Natural Products. Stockholm, Sweden Section 2C–1

Nasset ES (1938) Enterocrinin, a hormone which excites the glands of the small intestine. Am J Physiol 121:481–487

Nasset ES, Pierce HB, Murlin JR (1935) Proof of a humoral control of intestinal secretion. Am J Physiol 111:145–158

O'Dorisio TM, Cataland S, Stevenson M, Mazzaferri EL (1976) Gastric inhibitory polypeptide (GIP). Intestinal distribution and stimulation by amino acids and medium-chain triglycerides. Am J Dig Dis 21:761–765

O'Dorisio TM, Sirinek KR, Mazzaferri EL, Cataland S (1977) Renal effects on serum gastric inhibitory polypeptide (GIP). Metabolism 26:651–656

Pavlov IV (1910) The work of the digestive glands. Translated by Thompson WH. Griffin, London

Pearse AGE (1969) The cytochemistry and ultrastructure of polypeptide hormone-producing cells of the APUD series and the embryologic, physiologic and pathologic implications of the concept. J Histochem Cytochem 17:303–313

Pederson RA, Brown JC (1972) Inhibition of histamine-, pentagastrin-, and insulin-stimulated canine gastric secretion by pure 'gastric inhibitory polypeptide'. Gastroenterology 62:393–400
Pederson RA, Brown JC (1976) The insulinotropic action of gastric inhibitory polypeptide in the perfused isolated rat pancreas. Endocrinology 99:780–785
Pederson RA, Brown JC (1978) Interaction of gastric inhibitory polypeptide, glucose and arginine on insulin and glucagon secretion from the perfused rat pancreas. Endocrinology 103:610–615
Pederson RA, Dryburgh JR, Brown JC (1975a) The effect of somatostatin on release and insulinotropic action of gastric inhibitory polypeptide. Can J Physiol Pharmacol 53:1200–1205
Pederson RA, Schubert HE, Brown JC (1975b) Gastric inhibitory polypeptide. Its physiological release and insulinotropic action in the dog. Diabetes 24:1050–1056
Perley MJ, Kipnis DM (1966) Plasma insulin response to glucose and tolbutamide of normal weight and obese diabetic and nondiabetic subjects. Diabetes 15:867–874
Perley MJ, Kipnis DM (1967) Plasma insulin responses to oral and intravenous glucose: Studies in normal and diabetic subjects. J Clin Invest 46:1954–1962
Pfeiffer EF, Telib M, Ammon J, Melani F, Ditschuneit H (1965) Direkte Stimulierung der Insulinsekretion in vitro durch Sekretin. Dtsch Med Wochenschr 90:1663–1667
Pincus IJ, Thomas JE, Rehfuss ME (1942) A study of gastric secretion as influenced by changes in duodenal acidity. Proc Soc Exp Biol 51:367–368
Polak JM, Bloom SR, Kuzio M, Brown JC, Pearse AGE (1973) Cellular localization of gastric inhibitory polypeptide in the duodenum and jejunum. Gut 14:284–288
Polak JM, Pearse AGE, Grimelius L, Marks V (1975) Gastrointestinal apudosis in obese hyperglycaemic mice. Virchows Arch [Cell Pathol] 19:135–140
Preshaw RM, Grossman MI (1965) Comparison of subcutaneous and intravenous administration of pancreatic stimulants. Am J Physiol 209:803–810
Quigley JP, Phelps KR (1934) The mechanism of gastric motor inhibition from ingested carbohydrate. Am J Physiol 109:133–138
Rabinovitch A, Dupré J (1972) Insulinotropic and glucagonotropic activities in crude preparations of cholecystokinin-pancreozymin. Clin Res 20:945
Rabinowitz D, Zierler KL (1962) Forearm metabolism in obesity and its response to intraarterial insulin. Characterization of insulin resistance and evidence for adaptive hyperinsulinism. J Clin Invest 41:2173–2181
Ramaswamy K, Malathi P, Caspary WF, Crane RK (1974) Studies on the transport of glucose from disaccharides by hamster small intestine in vitro. II. Characteristics of the disaccharidase-related transport system. Biochem Biophys Acta 345:39–48
Raptis S, Rau RM, Schroder KE, Hartmann W, Faulhaber JD, Clodi PH, Pfeiffer EF (1971) The role of the exocrine pancreas in the stimulation of insulin secretion by intestinal hormones. III. Insulin responses to secretin and pancreozymin in experimentally induced pancreatic exocrine insufficiency. Diabetologia 7:68–72
Reichlin M, Schnure JJ, Vance VK (1968) Induction of antibodies to porcine ACTH in rabbits with non-steroidogenic polymers of BSA and ACTH. Proc Soc Exp Biol Med 128:347–350
Reynolds C, Tronsgard N, Gibbons E, Blix PM, Rubenstein AH (1979) Gastric inhibitory polypeptide response to hyper- and hypoglycaemia in insulin-dependent diabetics. J Clin Endocrinol Metab 49:255–261
Ross SA, Dupré J (1978) Effects of ingestion of triglyceride or galactose on secretion of gastric inhibitory polypeptide and on responses to intravenous glucose in normal and diabetic subjects. Diabetes 27:327–333
Ross SA, Brown JC, Dupré J (1977) Hypersecretion of gastric inhibitory polypeptide following oral glucose in diabetes mellitus. Diabetes 26:525–529
Sarson DL, Bryant MG, Bloom SR (1980) A radioimmunoassay of gastric inhibitory polypeptide in human plasma. J Endocrinol 85:487–496
Schäfer R, Schatz H (1979) Stimulation of (pro-)insulin biosynthesis and release by gastric inhibitory polypeptide in isolated islets of rat pancreas. Acta Endocrinol 91:493–500
Schauder P, Brown JC, Frerichs H, Creutzfeldt W (1975) Gastric inhibitory polypeptide: Effect on glucose-induced insulin release from isolated rat pancreatic islets in vitro. Diabetologia 11:483–484
Schwartz CK, Kimberg DV, Sheerin HE, Field M, Said SI (1974) Vasoactive intestinal peptide stimulation of adenylate cyclase and active electrolyte secretion in intestinal mucosa. J Clin Invest 54:536–544

Service FJ, Nelson RL, Rubinstein AH, Go VLW (1978) Direct effect of insulin on secretion of insulin, glucagon, gastric inhibitory polypeptide, and gastrin during maintenance of normoglycaemia, J Clin Endocrinol Metab 47:488–493

Shima K, Kuroda K, Matsuyama T, Tarui S, Nishikawa M (1972) Plasma glucagon and insulin responses to various sugars in gastrectomized and normal subjects. Proc Soc Exp Biol Med 139:1042–1048

Sircus W (1958) Studies on the mechanisms in the duodenum inhibiting gastric secretion. Q J Exp Physiol 43:114–133

Sirinek KR, Cataland S, O'Dorisio TM, Mazzaferri EL, Crockett SE, Pace WG (1977) Augmented gastric inhibitory polypeptide response to intraduodenal glucose by exogenous gastrin and cholecystokinin. Surgery 82:483–442

Sirinek KR, Pace WG, Crocket SE, O'Dorisio TM, Mazzaferri EL (1978) Insulin-induced attenuation of glucose-stimulated gastric inhibitory polypeptide secretion. Am J Surg 135:151–155

Smith PH, Merchant FW, Johnson DG, Fujimoto WY, Williams RH (1977) Immunocytochemical localization of gastric inhibitory polypeptide-like material within A-cells of the endocrine pancreas. Am J Anat 149:585–590

Solcia E, Capella C, Vasallo G, Buffa R (1974) Endocrine celles of the gastric mucosa. In Rev Cytol 42:223–286

Soon-Shiong P, Debas HT, Brown JC (1979a) The evaluation of GIP as an enterogastrone. J Surg Res 26:681–686

Soon-Shiong P, Debas HT, Brown JC (1979b) Cholinergic inhibition of gastric inhibitory polypeptide (GIP) action. Gastroenterology 76:1253

Sykes S, Morgan LM, English J, Marks V (1980) Evidence for preferential stimulation of gastric inhibitory polypeptide secretion in the rat by actively transported carbohydrates and their analogues. J Endocrinol 85:210–207

Thomas FB, Mazzaferri EL, Crockett SE, Mekhjian HS, Gruemer HD, Cataland S (1976) Stimulation of secretion of gastric inhibitory polypeptide and insulin by intraduodenal amino acid perfusion. Gastroenterology 70:523–527

Thomas FB, Shook, O'Dorisio TM, Cataland S, Mekhjian HS, Caldwell JH, Mazzaferri EL (1977) Localization of gastric inhibitory polypeptide release by intestinal glucose perfusion in man. Gastroenterology 72:49–54

Thomas FB, Sinar D, Mazzaferri EL, Cataland S, Mekhjian HS, Caldwell JH, Fromkes JJ (1978) Selective release of gastric inhibitory polypeptide by intraduodenal amino acid perfusion in man. Gastroenterology 74:1261–1265

Thomford NR, Sirinek KR, Crockett SE, Mazzaferri EL, Cataland S (1974) Gastric inhibitory polypeptide. Arch Surg 109:177–182

Unger RH, Eisentraut AM (1969) Entero-insular axis. Arch Intern Med 123:261–266

Unger RH, Ketterer H, Dupré J, Eisentraut AM (1967) The effects of secretin, pancreozymin and gastrin on insulin and qlucagon secretion in anesthetized dogs. J Clin Invest 46:1630–1642

Vagne M, Stening GF, Brooks FP, Grossman MI (1968) Synthetic secretin: Comparison with natural secretin for potency and spectrum of physiological actions. Gastroenterology 55:260–267

Vaitukaitis J, Robbins JB, Neuschlag E, Ross GT (1971) A method for producing specific antisera with small doses of immunogen. J Clin Endocrinol Metab 33:988–991

Vassallo G, Capella C, Solcia E (1971) Endocrine cells of the human gastric mucosa. Z Zellforsch 118:48–67

Villar HV, Fender HR, Rayford PL, Bloom SR, Ramus NI, Thompson JC (1976) Suppression of gastrin release and gastric secretion by gastric inhibitory polypeptide (GIP) and vasoactive intestinal polypeptide (VIP) Ann Surg 184:97–102

Vinik AI, Kalk WJ, Jackson WPU (1974) A unifying hypothesis for hereditary and acquired diabetes Lancet I:485–486

Wang CC, Grossman MI (1951) Physiological determination of release of secretin and pancreozymin from intestine of dog with transplanted pancreas. Am J Physiol 164:527–545

Way, LW (1970) Effect of secretin on gastric acid secretion in response to cholinergic stimuli. Gastroenterology 59:510–517

Willms B, Ebert R, Creutzfeldt W (1978) Gastric inhibitory polypeptide (GIP) and insulin in obesity: II. Reversal of increased responses to stimulation by starvation or food restriction. Diabetologia 14:379–387

Wormsley KG, Grossman MI (1964) Inhibition of gastric acid secretion by secretin and by exogenous acid in the duodenum. Gastroenterology 47:72–81

Wright JP, Barbezat GO, Clain JE (1979) Jejunal secretion in response to a duodenal mixed nutrient perfusion. Gastroenterology 76:94–98

Yajima H, Ogawa H, Kubota M, Tobe T, Fujimura M, Henmi K, Torizuka K, Adachi H, Imura H, Taminato T (1975) Synthesis of the tritetracontapeptide corresponding to the entire amino acid sequence of gastric inhibitory polypeptide. J Am Chem Soc 97:5593–5594

Yanaihara N, Mochizuki T, Yanaihara C, Sakura N, Hashimoto T, Sakagami M, Kaneko T, Kanetoh A, Brown JC (1978) Synthesis of GIP. In: Bloom SR (ed) Gut hormones. Churchill Livingstone, Edinburgh London New York, pp 271–276

Zunz E, Labarre J (1929) Contributions à l'étude des variations physiologiques de la secretion interne du pancreas: Relations entre les secretions externe et interne du pancreas. Arch Int Physiol 31:20–44

Acknowledgements

The author acknowledges with appreciation the following Journals and Publishers for permission to reproduce figures:

The American Diabetes Association Inc.
William Heinemann Medical Books Ltd.
Rockefeller University Press
British Medical Association
J.B.Lippincott Co.
The Williams and Wilkins Co.
Springer-Verlag GmbH & Co. KG
Academic Press Inc.
Charles B. Slack Inc.
S. Karger AG
Elsevier/North Holland
The American Physiology Society
The Scandinavian Journal of Gastroenterology

Subject Index

Other Volumes in This Series:

Volume 23: E. Flückiger, E. Del Pozo, K. v. Werder
Prolactin
Physiology, Pharmacology and Clinical Findings
1982. 54 figures, 14 tables. X, 224 pages. ISBN 3-540-11071-2

Volume 22: D. T. Krieger
Cushing's Syndrome
1982. 27 figures in 42 separate illustrations (some in color)
IX, 142 pages. ISBN 3-540-10811-4

Volume 21: A. E. Schindler
Hormones in Human Amniotic Fluid
1982. 23 figures, 133 tables. XII, 158 pages. ISBN 3-540-10810-6

Volume 20: R. Volpé
Auto-immunity in the Endocrine System
1981. 32 figures, 15 tables. X, 187 pages. ISBN 3-540-10677-4

Volume 19: P. Mauvais-Jarvis, F. Kuttenn, I. Mowszowicz
Hirsutism
1981. 32 figures, 10 tables. XI, 110 pages. ISBN 3-540-10509-3

Volume 18: I. J. Chopra
Triiodothyronines in Health and Disease
With a Contribution by V. Cody
1981. 76 figures, 18 tables. IX, 145 pages. ISBN 3-540-10400-3

Volume 17: J. Chayen
The Cytochemical Bioassay of Polypeptide Hormones
1980. 72 figures, 7 tables. XIV, 208 pages. ISBN 3-540-10040-7

Volume 16: J. E. A. McIntosh, R. P. McIntosh
Mathematical Modelling and Computers in Endocrinology
1980. 73 figures, 57 tables. XII, 337 pages. ISBN 3-540-09693-0

Volume 15: A. T. Cowie, I. A. Forsyth, I. C. Hart
Hormonal Control of Lactation
1980. 64 figures, 7 tables. XIV, 275 pages. ISBN 3-540-09680-9

Volume 14: J. H. Clark, E. J. Peck, Jr.
Female Sex Steroids
Receptors and Function
1979. 116 figures, 18 tables. XII, 245 pages. ISBN 3-540-09375-3

Volume 13: H. F. De Luca
Vitamin D – Metabolism and Function
1979. 14 figures. VIII, 80 pages. ISBN 3-540-09182-3

Springer-Verlag
Berlin
Heidelberg
New York

Volume 12:
Glucocorticoid Hormone Action
Editors: J.D.Baxter, G.G.Rousseau
1979. 176 figures, 58 tables. XIX, 638 pages. ISBN 3-540-08973-X

Volume 11: S.Ohno
Major Sex-Determining Genes
1979. 34 figures, 6 tables. XIII, 140 pages. ISBN 3-540-08965-9

Volume 10: W.I.P.Mainwaring
The Mechanism of Action of Androgens
1977. 12 figures, 17 tables. XI, 178 pages. ISBN 3-540-07941-6

Volume 9: R.E.Mancini
Immunologic Aspects of Testicular Function
1976. 36 figures, 8 tables. IX, 114 pages. ISBN 3-540-07496-1

Volume 8: E.Gurpide
Tracer Methods in Hormone Research
1975. 35 figures. XI, 188 pages. ISBN 3-540-07039-7

Volume 7: E.W.Horton
Prostaglandins
1972. 97 figures. XI, 197 pages. ISBN 3-540-05571-1

Volume 6: K.Federlin
Immunopathology of Insulin
Clinical and Experimental Studies
1971. 53 figures. XIII, 185 pages. ISBN 3-540-05408-1

Volume 5: J.Müller
Regulation of Aldosterone Biosynthesis
1971. 19 figures. VII, 137 pages. ISBN 3-540-05213-5

Volume 4: U.Westphal
Steroid-Protein Interactions
1971. 144 figures. XIII, 567 pages. ISBN 3-540-05313-3

Volume 3: F.G.Sulman
Hypothalamic Control of Lactation
In collaboration with numerous experts.
1970. 58 figures. XII, 235 pages. ISBN 3-540-04973-8

Volume 2: K.B.Eik-Nes, E.C.Horning
Gas Phase Chromatography of Steroids
1968. 85 figures. XV, 382 pages. ISBN 3-540-04277-6

Volume 1: S.Ohno
Sex Chromosomes and Sex-linked Genes
1967. 33 figures. X, 192 pages. ISBN 3-540-03934-1

Springer-Verlag
Berlin
Heidelberg
New York